INSIDE THE WORM

ROBERT SWINDELLS

Illustrated by Jon Riley

INSIDE THE WORM

A YEARLING BOOK 0 440 86300 7

First published in Great Britain by Doubleday, a division of
Transworld Publishers Ltd.

PRINTING HISTORY
Doubleday edition published 1993
Yearling edition published 1994

Yearling Books are published by Transworld Publishers Ltd,
61–63 Uxbridge Road, Ealing, London W5 5SA,
in Australia by Transworld Publishers (Australia) Pty. Ltd,
15–25 Helles Avenue, Moorebank, NSW 2170,
and in New Zealand by Transworld Publishers (NZ) Ltd,
3 William Pickering Drive, Albany, Auckland.

Printed and bound in Great Britain by
Cox & Wyman Ltd, Reading, Berks.

The worm danced sinuously through the darkening garage, its great head swaying and bobbing. Now and then its reflector eyes would catch light from somewhere and flash red. Fliss was amazed at the dexterity of her friends; their co-ordination. The way their dancing feet avoided the great train of fabric they trailed, which slid, hissing, across the dusty concrete. The ease with which they seemed to have mastered the technique. Their shouts of laughter grew louder as Gary increased his speed, but there were no disasters – nobody stumbled. Fliss watched as though mesmerized, and when she remembered to look at her watch it was ten to ten.

'Hey!' Their exultant laughter drowned her voice. 'Hey, you guys. It's almost ten. I've got to go.'

Nobody heard. Gary shifted up another gear and they came whooping in his wake, precisely, like a well-drilled squad.

'Lisa?' Surely her best friend would respond – break step so that the dance could end in red-faced, panting laughter?

'Lisa?' they mimicked, and her voice was among them. 'Lisa, Lisa, Lisa, Lisa –' The worm was coming at her now, eyes burning, jaws agape.

She turned and fled . . .

Also available by Robert Swindells,
and published by Yearling Books:

ROOM 13
Winner of the 1990 Children's Book Award

DRACULA'S CASTLE

HYDRA

THE POSTBOX MYSTERY

THE THOUSAND EYES OF NIGHT

For Nan
fighting her dragon

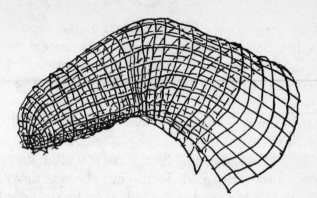

CHAPTER ONE

Fliss stuck her hand up. 'Why's it called a worm if it was a dragon, Sir?'

Mr Hepworth nodded. 'Good point, Felicity. To us today, the word "worm" conjures up a picture of a small, pink, harmless creature, doesn't it? But in Anglo-Saxon times, dragons and other reptilian monsters were often called worms, so the word would have had pretty dreadful connotations for them. The worm which terrorized Elsworth is said to have been a chain in length and five feet high at the shoulder.'

'How long's a chain, Sir?' asked Grant Cooper.

'Sixty-six feet. That's roughly twenty-two metres.'

'Phew – some worm!'

7

'Well yes, exactly. And five feet at the shoulder – that's like a fairly big horse, and then there'd be the neck and head, so we're not talking about something you could chop in bits with a garden spade.' The class tittered.

'And Saint Ceridwen went out by herself to face it, Sir?'

'Yes, Marie, she did. She wasn't a saint then, of course – just a village maiden – but she was devoutly religious and believed that God would empower her to overcome the worm, which she called an agent of Satan.'

'I wouldn't have gone, Sir.'

'No, Marie, and neither would I. We don't have Ceridwen's faith, you see.'

'Is it a true story, Sir? I mean, I thought there were no such things as dragons.'

The teacher smiled. 'There are no dragons now, Marie, but this was a thousand years ago, so who knows? Ceridwen certainly existed, and she must have done something pretty remarkable because we know she'd become the most important person in the district by the time she was martyred by the Danes in nine ninety-three.'

David Trotter raised his hand. 'How did she kill the worm, Sir?'

'She didn't. According to the legend, the moment the beast touched the hem of Ceridwen's skirt it became docile, whereupon she commanded it to begone. It slunk away on to the fen and was never seen again.'

Gary Bazzard grinned. 'It might still be out there, Sir.'

'I doubt it, Gary. Elsworth's got you now – it doesn't need another monster.'

'Sir,' said Fliss. 'Why did the Danes kill Ceridwen?'

The teacher shrugged. 'The Danes were pagans, Felicity. When they overran this area they demanded that Ceridwen worship their gods and order her people to do the same. She refused, so they hacked off her limbs and beheaded her.'

'Ugh! And this was exactly a thousand years ago, and that's why the town's having this Festival?'

Mr Hepworth nodded. 'That's correct, and the vicar of St Ceridwen's has invited our school to be involved in various ways, and Mrs Evans and I decided we'd ask Year Eight to perform a re-enactment of Ceridwen's encounter with the worm, and of her martyrdom. It's a great honour – the eyes of the whole town will be on us, so obviously we want to make a first-class job of it and it's all got to be ready in three weeks.'

'So it'll be sort of like doing a play, Sir?'

'That's right, Maureen, and the idea is that you people take a lot of the responsibility yourselves for producing it. Mrs Evans and I will be around if you need us, of course, but we expect you to write a script, do the casting, see to props and costumes and so forth. I think you'll enjoy the experience, but I want you to remember at all times that your finished effort will be seen by practically everybody in Elsworth, so the reputation of Bottomtop Middle is in your hands. That's all, I think. You can go now, and start work as soon as you like.'

'I'm playing the Boss Viking!' cried Gary Bazzard, as Year Eight spilled on to the playground. 'And I'll hack the limbs off anyone who argues.'

Fliss pulled a face at Lisa. 'Old Hepworth must be mad, putting the school's reputation in the hands of guys like him.'

Lisa laughed. 'Gary's a loudmouth, but he's OK. We can always gang up on him – tell him we've got him down to play Ceridwen in a blonde wig and a long white dress.'

That night, Fliss had a dream. In her dream the worm came slithering out of the fenland mist with a thousand-year hunger in its belly and vengeance in its brain and she, cast as Ceridwen by the votes of all her friends, was sent to stand defenceless in its path.

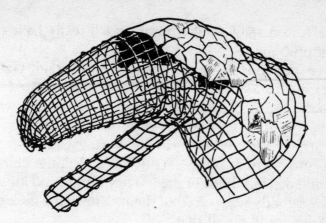

CHAPTER TWO

They'd learned about the play on Monday. Lunchtime Tuesday there was a class meeting to get the thing off the ground. No teachers were present, though Mrs Evans kept buzzing in and out because they were using her room.

'Right. Now – first things first.' Sarah-Jane Potts, who'd done some acting with a local amateur group, seemed to be chairing the meeting. 'Where will this play be performed?'

'On the Festival Field,' said Tara Matejak. 'Mr Hepworth said so.'

'So the audience will be all round us and there'll be some noise as well. That affects how we arrange

11

ourselves on stage, and it means we'll really have to speak up.'

'I've got this very powerful voice,' said Gary Bazzard. 'It's a Viking Chief's voice, really.'

Sarah-Jane had been tipped off by Lisa and was ready for him. 'Ah well, you can just forget it, Bazzard. We're having a girl for Viking Chief.'

'A girl?' cried Gary. 'You can't. Viking chiefs commanded hundreds of men. They fought and killed and everybody was scared of them. You name me one girl who could do all that.'

'How about Boudicca?'

'Who?'

'Boudicca, queen of the Iceni. She led an army against the Romans. And then there's Cartimandua, queen of the Brigante. She fought the Romans too.'

'You're making it up. You'll be telling me next that Arnold Schwarzenegger never goes anywhere without his knitting.'

'I'm not, and I won't,' retorted Sarah-Jane. 'But the Viking Chief's a girl, and that's that.'

'Which girl?' Gary wasn't about to give up.

'We don't know. We haven't voted yet.'

When they did vote, Gemma got the part. As the result was being announced, Mrs Evans walked in. 'Don't forget your understudies, Sarah-Jane,' she said.

'No way, Miss,' said Sarah-Jane, though she had forgotten.

'What're understudies?' asked Barry Tune.

Mrs Evans smiled. 'An understudy is someone who learns the part of a leading actor or actress, so that he or she can step in and play the part if the star falls ill. It's important to have understudies for all your leading roles – any teacher will tell you that.' She found the book she'd come for and left the room. The class then voted, and Maureen O'Connor was chosen as understudy.

And so it went on. There was consolation for Gary when the class made him worm's head. 'You get to roar, bare your fangs and breathe fire,' Lisa told him. 'What more could anyone ask?'

'If Gary's part of the worm, I want to be in it too,' said David Trotter. He and Gary were best friends. There would be four people in the worm, but there was no voting except for the head. Trot's offer was accepted, and Ellie-May Sunderland and Lisa got the two remaining places. Fliss landed the best part of all, beating Samantha by one vote to play Ceridwen, with Samantha as understudy.

After the allocation of supporting roles, Year Eight turned its attention to the problem of costume. It was decided that people would be responsible for designing and making their own costumes, though some children who were good at sewing would stand by to help if needed. The worm must be twenty-two metres long, and light enough for four

13

people to operate. 'Trouble is, it'll have eight legs,' said Gary. 'Real dragons have four.'

'How the heck do you know?' demanded Fliss. 'Have you seen one?'

'I've seen pictures.'

Fliss snorted. 'I've seen pictures of women with six arms,' she said. 'Doesn't mean women're like that, does it?'

By the time the meeting ended, everybody had something to do. They even had a title for their play: *Ceridwen – Heroine-Saint of Elsworth*. Robert Field had thought it up and everybody liked it. As she walked with Lisa to their own room for register, Fliss felt they'd made a really good start. She hadn't forgotten her nightmare, but in the warm light of afternoon a dream is just a dream.

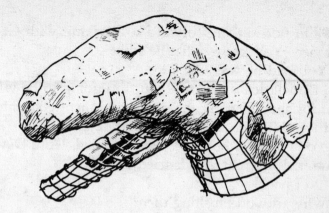

CHAPTER THREE

'How long did old Hepworth say we'd got?' asked Lisa, as she and Fliss walked home that afternoon.

'Three weeks, wasn't it?' Fliss began calculating aloud. 'We're in the first week of April, right? Festival week starts on Saturday the twenty-fourth and our play's the following Saturday, which is May the first. So we've got about three weeks by my reckoning. Why?'

'Oh, I was just wondering. There's a lot to do, isn't there?'

Fliss shrugged. 'Costume to make, lines to learn. It won't take all that much doing. You don't even need a costume – they'll only see your legs.'

'I know, but I've got to help with the worm,

15

and I'm not looking forward to working with Gary Bazzard. You know what he's like.'

'You said he was OK.'

'In small doses he's OK, but I'm going to be with him for ages, making the worm and then rehearsing, and I won't even have you to talk to.'

'You'll have Ellie-May.' Fliss grinned. 'And David Trotter. I thought you fancied Trot?'

'Do I heck!'

'Why are you blushing then?'

'I'm not.'

'Oh, I thought you were. Anyway, I'll tell you what.'

'What?'

'If you like, and if the others'll let me, I'll help with the worm.' She smiled. 'My costume's already made, you see.'

Lisa looked at her. 'How d'you manage that?'

'Well, all I need is a long white dress, and I've got one from when I was bridesmaid to my cousin last year. I've been dying for an excuse to wear it.'

'And you'll really come and work on the worm with me?'

'If it's all right with the others, yes.'

'That'll be great, Fliss. We're doing it at Trot's place, in his dad's garage. Apparently there's loads of junk there we can use – wire and old curtains and stuff. Trot says the worm's going to look like

one of those dancing lions they have in Thailand –
you'll have seen 'em on telly.'

'Yes, I have. I think it's a good idea, but ours'll
need a fiercer head. Thai lions don't look scary at all
– they're cute and cuddly.' She grinned. 'Like Trot.'

'Shut up.' Lisa kicked a stone into the verge as
her cheeks flamed. 'I can't stand him, if you must
know.'

'Why did you volunteer for the worm, then?'

'Shut up, Fliss, OK?'

Her friend chuckled. 'OK. When's the first session,
Lisa?'

'Tonight. Half-six. You coming?'

'Dunno, do I? Depends how Trot feels really – it's
his place. Phone him, then phone me. If he agrees,
I'll be there.'

Their ways parted soon after that and Fliss hurried
home. It'll be great, she told herself, working with
Lisa and the others: creating the monster I'll face on
the Festival Field.

So why did I shiver just now?

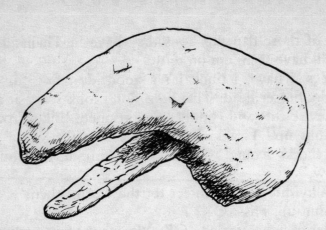

CHAPTER FOUR

Lisa waited till her watch showed one minute past six, then picked up the phone. Her mum was always telling her it was cheaper after six. She punched in Trot's number, feeling once more the slight tingle of excitement she always got when she did this. It's not true what Fliss says, she told herself. About me and Trot. I like him, that's all. We're friends.

There was a click and Trot's voice said, 'Elsworth four-six-four-two-six-two.'

'Trot? It's Lisa. Listen. Is it all right if Fliss comes tonight? She offered to help with the worm and I said I'd ask you.'

' 'Course it is. Many hands make light work, as my dad would say. Half-six, right?'

'Half-six. See you.' She broke contact and punched in Fliss's number. 'Fliss? Oh, sorry Mrs Morgan, it's Lisa. May I speak to Fliss, please? Thanks. Fliss? Lisa. I called Trot. It's OK for tonight.'

'Great. See you in twenty-five minutes then.'

'Right. 'Bye.'

'I did a rough design,' said Trot, unfolding a sheet of paper.

The four gathered round to see. Mr Trotter had backed his car on to the driveway so they'd have plenty of space.

Ellie-May frowned. 'It looks like a ladder.'

Trot nodded. 'I know, except the rungs are too far apart. This is the basic framework, see? We'd stand in a line with our heads between the rungs and the shafts resting on our shoulders. These hoops,' he pointed, 'are made of wire. They'd run from one shaft to the other like a series of arches, supporting the fabric covering well above our heads and giving the worm's back a nice rounded shape.'

Lisa nodded. 'You're a genius, Trot. It's brilliant.'

Gary nodded. 'Looks sound to me, man. Where do we get the stuff to make it?'

'It's all here.' Trot nodded to where some lengths of timber stood propped in a corner. 'There're the shafts, and we can make rungs from that too. Dad got it to build a porch and never got round to it.

19

And there's a coil of wire for the hoops.'

'What about nails?' asked Fliss.

'Drawerful in the chest there,' Trot told her. 'Staples too, for the hoops. We can have the framework done tonight if we get a move on.'

They did better than that, working together smoothly so that by eight o'clock they had a sturdy framework four metres long and almost a metre in height. They stood, fists on hips, looking at it. 'Four metres,' grunted Gary. 'The real worm was twenty-two.'

'Oh sure,' agreed Trot, 'but a framework that length would be so unwieldy we wouldn't be able to shift it. No, the rest'll be made up of neck and head, and a tail of fabric stiffened with wire.' He laughed. 'What we'll use for the head I don't know.'

'Papier-mâché,' said Fliss. 'It's light, and you can mould it into any shape you want.'

'Take a lot of paper,' said Ellie-May.

'Well, there's five of us,' said Lisa. 'If we get all the newspapers from home and from relatives, we'll have plenty.'

Trot nodded. 'Papier-mâché it is, then. Shall we meet here tomorrow, same time, to make a start?'

Lisa hung around when the others left. It wasn't fair to leave Trot with all the clearing away, and in any case she felt like walking home alone.

20

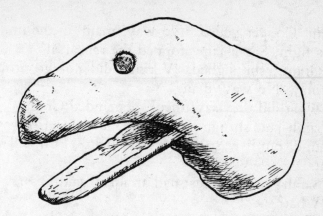

CHAPTER FIVE

The rest of that week was a busy one for Year Eight. Every spare minute of the school day was spent in discussing the play, and in the evenings the children worked on their costumes. Everybody had a part, as either a villager or a Viking, and the homes of aunts and grandparents were ransacked for materials which might do for a tunic, a helmet or a long dress.

On Thursday afternoon they gave up double games to stage a rough rehearsal on the school field. There were no written parts, so everybody had to make up their lines as they went along. Ad libbing, Sarah-Jane called it, but it wasn't a success. It's not easy thinking up the right words instantly, and when the Viking

21

Grant Cooper yelled, 'No way, man!' in the middle of a fight, Sarah-Jane stopped the rehearsal.

'Grant,' she sighed, 'Vikings did not go around saying, "No way, man." '

'What did they say, then?' demanded Grant.

Sarah-Jane shrugged. 'I don't know, do I? I wasn't around, but it wasn't "No way, man", I can tell you that.'

'Sarah-Jane, I've just had an idea,' said Fliss.

'What?'

'Well, we don't know how people spoke in those days, do we? Nobody does. So why don't we do it without words?'

Sarah-Jane looked at her. 'You mean mime it, or do it through dance or something?'

Fliss shook her head. 'No. I thought we could have a narrator. You know – somebody who stands at the side and tells the story as the play unfolds. That way, nobody has to learn lines and we can concentrate on the action.'

'The narrator'd have a lot to memorize.'

'Not necessarily. He or she could read from a script done up to look like an ancient chronicle or something. Nobody'd be watching the narrator anyway, if we made the action exciting enough.'

'Hmmm.' Sarah-Jane frowned. 'It's an idea, Fliss. It'd get rid of "No way, man," and stuff like that, but who's going to do it?'

'I will,' volunteered Andrew Roberts, 'if someone'll help me write it.'

'We'll all help to write it,' smiled Sarah-Jane. 'Thanks, Andrew.'

For the moment they carried on with no words except those of Sarah-Jane, who was directing. They hit another snag after Ceridwen banished the worm. 'What do we do now?' asked Barry Tune. 'I mean, years go by before the Danes come and kill her.'

'Hmmm.' Sarah-Jane frowned again.

'We could have a ceremony,' suggested Waseem. 'You know – the villagers are so grateful to Ceridwen they make her their chief or something.'

'Yes,' put in Haley. 'And remember, the Vikings were raiding long before they settled here. We could show a series of unsuccessful raids with the Danes being repulsed by the villagers.'

Sarah-Jane nodded. 'Good idea, Haley. Yours too, Waseem. Let's try it.'

They tried it, and it worked. Friday lunchtime they went through it again, this time in the hall. The first bit of narration was ready and Andrew read it as they performed. It looked good. 'All we need now is the costumes,' grinned Marie. 'And we're the Royal Shakespeare Company.'

CHAPTER SIX

Saturday morning Fliss had to go with her mother to buy shoes. Then there was lunch, and by the time she got to Trot's garage the others had practically finished the head. She gasped when she saw it. It was enormous, and looked fantastic with its red eyes and gaping jaws. 'Wow!' she cried. 'Those eyes are really ace, Trot. What're they made of?'

Trot grinned. 'Reflectors, Fliss, from Gary's dad's old car. D'you like 'em?'

'Like 'em? They're amazing. It's like they're staring right at you, hating you. What a terrific idea.'

'Yeah, well – we need a terrific idea from you, Fliss, now that you've finally shown up.'

'Why – what's up?'

'It's the neck,' said Lisa. 'It's designed to go over Gary's head and shoulders and down to his waist, so that the head is firmly supported and won't sway about when the worm's moving.'

'And doesn't it fit?'

'Oh, it fits all right, but it pins Gary's arms to his sides. He feels like an Egyptian mummy in there and it's not safe for him to walk, let alone run. If he tripped, he'd fall flat on his face.'

'Hmmm.' Fliss looked at the head. 'Is the papier-mâché completely dry now?'

'The thickest parts are still a bit soggy, but it's OK. Why?'

'Well, if the neck's dry we could take a saw and cut two slots in it, one either side. It'd still reach his waist back and front, and his arms would be free.'

'Fliss Morgan, you're a genius,' cried Trot. 'An infant prodigy. Why didn't we think of that?'

The slots were quickly cut, and Gary tried on the head. He couldn't see yet because they hadn't made the eye-holes, but they led him on a circuit of the garage and he did some roaring and said he felt much better. Now that the papier-mâché had dried out, the whole thing was surprisingly light. They spent the afternoon painting it, and by half-past four the last scrap of newsprint was covered and the head was a glossy green, except for the inside of the mouth which they'd done with some obscenely pink stuff Ellie-May

25

had got from somewhere. They propped it in a corner and stood in a half-circle, looking at it.

'It looks like a pensioner yawning,' said Lisa. 'It's got no teeth.'

'Don't worry,' said Ellie-May. 'My gran's got some things we can use for teeth.'

'What sort of things?'

'Oh – they're cone-shaped plastic things from where she used to work. Bobbins of some sort, I think. They're all colours, but we can soon paint 'em white.'

Trot looked at her. 'Can you bring them to-morrow?'

'No problem.'

'Right.' He turned to the others. 'Half-past ten then, here?'

This time, Lisa left with Fliss. Fliss grinned. 'Trot found somebody else, has he?'

Lisa shook her head. 'I told you – I don't care about Trot. I care about the play, that's all. I get a funny feeling every time I think about it.'

'What sort of feeling? Are you nervous?'

Lisa shook her head. 'Not nervous exactly. Sort of shivery. It's ever since we started the head.'

Fliss laughed. 'You scared of it?'

'Me? No. I don't need to be, Fliss. It's you. You're Ceridwen.'

Fliss pulled a face. 'I know. I had a nightmare. But

it's only a story, so there can't really be anything to be afraid of, can there?'

Her friend shrugged. 'I dunno. Maybe not. Anyway, can we talk about something else now, Fliss?'

CHAPTER SEVEN

Sunday morning was dull and drizzly, but Ellie-May had brought the teeth. Each tooth was twenty centimetres long and came to a good sharp point at one end. Everybody had come in old clothes and they spent a happy hour with the white paint, slapping it on the cones and standing them in a row on Trot's dad's workbench. When the last tooth was done, Trot counted them. 'Twenty-eight,' he said. 'Just right. Seven each side, top and bottom.'

'How do we fix 'em in?' asked Gary.

'Superglue,' Trot told him. 'We gouge out sockets in the papier-mâché, smear 'em with superglue and stick the teeth in. Nothing'll shift 'em once that glue sets. Nothing. But the paint's got to dry first.'

They made the sockets while they were waiting. It wasn't easy. The painted papier-mâché was remarkably tough. By the time they'd finished it was nearly lunchtime and the teeth were almost dry. 'Near enough, anyway,' said Trot, testing one with his finger. 'We can always touch 'em up after if they get fingerprints on 'em.'

By one o'clock the worm's head had a full set of fearsome teeth. The difference was amazing. 'Wow!' breathed Ellie-May. 'Look at it. It's so realistic.'

Trot nodded. 'Sure is. I mean, I know we wanted it scary, but this is almost too frightening. I'd have a fit if I met that in the woods at night.'

'Hey!' Gary's eyes shone. 'Help me on with it – let's see what it looks like moving.'

'No!' Lisa shook her head. 'Don't, Gary. Don't put it on.' She sounded frightened and everybody looked at her.

'What's the matter?' demanded Gary. 'You scared or something?'

Lisa nodded. 'Yes, I'm scared. I don't know why, but I am. It's too good. Too real. I can't believe we made it.'

Trot laughed. 'Who made it if we didn't, Lisa? We're geniuses, kid. It looks good because it was created by a team of brilliant minds. Come on, Gary – let's see our stunning creation in action.'

Lisa was backing towards the big double doors. 'I – I've got to go,' she murmured. 'Lunch. I'll see you at school tomorrow, OK?'

'Hey, Lisa.' Fliss looked at her friend. 'Hang on one minute, right? One minute and I'll be with you.'

'No, sorry.' Lisa's face was chalk white. 'I can't. I don't feel well. I have to go now.'

'Oh, all right, I'll come with you.' Fliss shot the others an apologetic glance. 'I'll see you tomorrow.'

When she got outside, Lisa was halfway down the drive. Fliss had to run to catch up. 'You're acting crazy, d'you know that?' she panted. 'They'll drop you from the team if you're not careful.'

'Let them,' said Lisa. 'I'll probably drop out anyway.'

'But why, Lisa? You volunteered for the worm, you know. Nobody forced you.'

'Listen.' Lisa spun on her heel and faced her friend. 'Don't you feel anything when you look at that head? Those teeth? I do. I feel like – like it's all too easy. I mean, that head's perfect, Fliss. Perfect. Stuff you make out of papier-mâché just doesn't turn out like that, especially when a lot of people work on it. You get lumps and dents – it ends up the wrong shape. You know what I mean.'

'Yes.' Fliss nodded. 'We do seem to have been lucky. We slapped the thing together and by sheer

30

chance it came out right. But does that mean we should be scared of it?'

'I'm not scared of it.'

'Well, you could have fooled me. You looked terrified, backing out of that garage. Look.' She put an arm round Lisa and squeezed her skinny waist. 'I'm your friend, right? Whatever's bugging you, you can tell me.'

Lisa nodded. 'I know.' They walked on. 'It's hard to explain, Fliss, but I'm not afraid of the worm. Not in the way you mean, but all the same there's something about it that's not quite right.' She smiled wanly. 'And if I still feel the same when it's finished, I think I'm going to have to drop out.'

'Well,' sighed Fliss, 'It is odd, I suppose, the way everything's come together so perfectly. Anyway, I'm on your side whatever happens. D'you want to meet up after lunch – walk round town or something?'

Lisa nodded. 'OK. Half-two?'

'Half-two, on the corner. I'll wear my new black jeans.'

CHAPTER EIGHT

'Hey, I like the gear!' cried Lisa. Fliss did a twirl, showing off her black jeans and top, her brand-new trainers.

'You've perked up a bit,' she grinned. 'Must've been hunger.'

Lisa smiled. 'Maybe.' They fell into step, strolling towards the town centre.

'What we gonna do?' asked Lisa.

Fliss shrugged. 'Not a lot. Everything'll be shut, but some of the kids could be around.' She looked sidelong at her friend. 'We might see Trot.'

'Shut up.'

Fliss was partly right. Most places were shut. McDonald's was open though, and they peered through the big window, looking for friends among the diners. There were none. 'We could go in anyway,' suggested Fliss. 'Have coffee and pie or something.'

'Ugh.' Lisa pulled a face. 'Do you mind? I've just eaten about half a cow and a truckful of veg. Haven't you had lunch? You must have a stomach like a Hoover.'

'I thought it'd be something to do, that's all.'

'I'd rather walk on, boring though it is.'

'I'll tell you what.'

'What?'

'It must've been even more boring when it was just a village.'

Lisa looked at her. 'What made you think of that?'

'The play, of course.'

'Hmmm.' Lisa grinned. 'No McDonald's, that's for sure, but they did have the worm to liven things up.'

Fliss giggled. 'D'you reckon it was exciting waiting for it to come out of the marsh, never knowing when it might be your turn to get eaten?'

Lisa shook her head. 'Horrible, I should think. Terrifying. Like a village in India when there's a man-eating tiger about. Not boring though.'

'I think I'd rather be bored.'

Lisa laughed. 'I'd rather be the worm.'

Fliss looked at her. 'Fancy human flesh, do you?'

Lisa chuckled. 'Not the flesh, Fliss. The power.'

'How d'you mean?'

'Well, think about it.' Lisa's eyes gleamed. 'Everyone running, hiding, shaking with terror every time you appeared. You could have anything you wanted – make them do whatever you wanted them to do. What a fantastic feeling that'd be.'

Fliss shook her head. 'I think I'd rather be liked, Lisa.'

Lisa laughed. 'That's the whole point, Fliss – you don't need to be liked if you're feared. If they're all scared stiff of you, they'll fall over each other to be your friend.'

There was a note in Lisa's voice which Fliss had never heard before. She gazed at her friend. 'Sounds like you've thought it all out, Lisa. I never knew stuff like that went on inside your skull.'

Lisa frowned. 'It didn't. Not till this worm thing started. It's doing my head in if you want to know, Fliss. I can't stop thinking about it.'

'Well, if it's bothering you so much, maybe you should drop out, but I don't get it. It's only a play, for Pete's sake, with a papier-mâché monster we made ourselves.' Fliss wished she felt as certain as she sounded.

Lisa nodded. 'I know, and I don't understand either. I—' She broke off, peering along the street they'd just turned into. Some way down, a great yellow skip stood on the pavement. Men were hurrying in and out of a building, throwing things into the skip, going back for more. 'What's going on, Fliss – what place is that?'

Fliss shrugged. 'Demolition, by the look of it. It's the Odeon.' The Odeon was Elsworth's only cinema. 'Come on.' She plucked at Lisa's sleeve. 'Let's watch for a bit. It'll be something to do, if nothing else.'

CHAPTER NINE

They watched from across the street. Two men brought out rolls of carpet and threw them in the skip. A great battered van came nosing along the street and drew up, blocking their view. They crossed over a bit further along and watched from there. The men were carrying out seats of steel and worn plush. These didn't go into the skip. They were lifted from the men's shoulders by two youths in the back of the van, who stacked them in the cavernous interior. 'Wonder where they're going?' whispered Fliss.

'Another cinema?' suggested Lisa.

When the van was crammed with seats, the two youths jumped down, secured its shutter-door and clambered into the cab. The engine coughed and roared and the van lurched away in a fog of blue exhaust. Three demolition men in vests and jeans stood, hands on hips, watching it leave. As they turned to go back inside, one of them noticed the two girls and called to them. 'Wanna buy a cinema, ladies?'

Lisa shook her head. 'Not today, thanks.'

'How much?' asked Fliss.

'Oh, let me see.' The man pretended to calculate. 'Five quid?'

'Sorry, haven't got it. How about sixty-two pence?'

'No chance. Good, solid building this, shoved up in nineteen thirty-two.'

'No seats in it though,' grinned Fliss.

'Yeah, there is – hundreds yet. And anyway, what d'you expect for five quid?'

'Sixty-two pence,' Fliss reminded.

'Aaa – miser, that's all you are.' He turned to follow his mates inside, muttering, 'Sixty-two pence!' as he went.

'You aren't half cheeky, Fliss,' giggled Lisa.

'No I'm not. He started it.'

'I wonder what he'd have done if you'd pulled a fiver out and said "OK"!'

'Sold it to me, of course.' Fliss chuckled. 'Can you imagine my mum's face if I walked in and said, "Mum, I bought the Odeon."?'

Lisa nodded. 'She'd say, "Well, you can't have

the dirty old thing in your room – you must keep it in the shed." '

The two friends were laughing so much when the man reappeared that he had to whistle piercingly to attract their attention. 'If I can't sell you a picture house, what about a nice bit of dress material?' They looked, and saw that he was carrying a mass of satiny green fabric which lay in shimmering folds across his arms. It was so slippery, and there was so much of it, that he was having to steady it with his chin to stop it toppling forward.

'Hey!' breathed Fliss. 'What is it?'

'Curtain,' the man told her. 'You know – they used to pull it across the screen between films. There's another just like it inside. D'you fancy it?'

'You bet. How much?'

The man laughed. 'I don't want your money, love. Here – take it. It'll only get burned if you don't.'

Fliss started forward, but Lisa's fingers snatched at her sleeve. 'What d'you want with an old curtain, Fliss? Come on – let's go, huh?'

'No.' Fliss freed her arm. 'Don't you see? It's exactly what we need to cover the worm with. It's green, it's shiny and it's very, very long. In other words, it's perfect.' She walked up to the demolition man.

'Watch it, love,' he grinned. 'It's heavy.' He tipped it into her arms, and she staggered under its weight.

Her knees buckled as she bore her prize back to Lisa.

'See?' she beamed. 'What a fantastic stroke of luck.'

Lisa shook her head. 'Not luck, Fliss. Fate.'

'What do you mean, fate? What are you on about, Lisa?'

'Fate is what I'm on about, Fliss. The thing that made the frame perfect and the head perfect and the teeth perfect. The thing that made us walk down here, today of all days, so that you could find a perfect skin for our perfect worm. Don't you see? It's all coming too easily.'

Fliss gazed into her friend's troubled eyes. 'Oh, Lisa – it's a run of luck, that's all. It happens. Are you going to help me carry this, or do I have to cripple myself?'

Lisa shrugged. 'I'll help. You know I will, but I wish we hadn't come this way. I wish the horrid thing was all burned up.'

CHAPTER TEN

'Wanna buy a powerhouse, ladies?'

She and Fliss were on a street. The man wore jeans and a vest and had perfect teeth. 'How much?' asked Fliss.

'McDonald's,' said the man. 'Shoved up in thirteen ninety-two.'

Fliss laughed. 'What d'you want with a powerhouse, Lisa? Lisa-pisa monkey-greaser.'

'Not the flesh, Fliss. The power. I've thought it all out.'

'I never knew stuff like that went on inside your skull. It's only papier-mâché.'

'Not to me it isn't. It's too perfect.'

'It's a run of luck, that's all.'

The man whistled shrilly to attract their attention. He was much further away now. 'You want this stuff or not?' He was cradling something in his arms but she couldn't see what it was.

'You bet!' she cried, running towards him. Behind her, Fliss was laughing. Her laughter echoed in the street.

For a long time the distance between the man and herself seemed to stay the same, and then in a moment she was with him. He smiled. Close up, she could see fingerprints on his teeth. 'We can always touch 'em up after,' he said, and she saw he'd turned into Trot. 'Here.' He held out what he was carrying.

Fear seized her. 'What is it?'

'A man-eating tiger. They used to pull it across the screen between films.'

'I – I don't like it.' She tried to back away but her feet wouldn't move.

'You don't need to be liked if you're feared.' He tipped the great, snarling cat into her arms and Lisa woke, screaming.

CHAPTER ELEVEN

'Are you sure you're all right to go this morning, Lisa?' Mrs Watmough eyed her daughter anxiously. She was pale, and the skin under her eyes looked bruised and puffy.

'Sure I am, Mum. I had a dream, that's all.'

'A nightmare, more like. You haven't screamed in the night like that since you were four.'

'I'm OK, Mum, honestly. I feel fine.' She didn't, but both her parents went out to work and she wasn't going to stay alone in the house all day.

'Well, if you're determined—'

''Bye, Mum.'

Determined. Lisa smiled faintly to herself as she walked down the path. I wish I was determined.

About anything. Confused is what I am. Mixed up. Scared, if you want to know the truth. Something's happening to me and I don't understand what it is, except that it's got something to do with the play. Well – we're due to meet with old Hepworth this aft to discuss progress. Maybe I'll ask to drop out. Dunno what excuse I'll come up with though – can't tell him I'm scared, can I?

Fliss was waiting for her at the top of the school drive. 'Hey, Lisa, you look awful. Is something wrong?'

'No, why should there be?'

Fliss shrugged. 'No reason.' She grinned. 'Anyway, here's Trot. He'll cheer you up.'

'Huh – fat chance.'

'Hi, girls,' Trot greeted. 'Seen Gary?'

They shook their heads. 'We've some good news for you though,' said Fliss.

'Let me guess – the school burned down?'

'No.'

'Old Hepworth's got measles?'

'Shut up and listen, will you? We've found a skin for the worm.'

'No kidding! What's it like?' Fliss described the material. 'Is there enough of it, though?'

Fliss nodded. 'It's a cinema curtain. It's higher than the school and nearly as long. It'd do for two worms.'

'Fantastic. Bring it to the garage tonight. Half-seven?'

'Right.'

Trot turned to Lisa. 'Half-seven OK for you?'

She pulled a face. 'Dunno. I might not come. Mum says I need an early night.'

Trot laughed. 'It's not a party, kid. No crates of booze. No rock band. You can be home by nine if that's what you want.'

'I don't know, Trot. I'll have to see, OK?' A part of her wanted to be there. The part that liked to be with Trot. But then there was that other part – the voice inside her head which was telling her to pull back – and that voice was growing louder.

'Sure.' Trot shrugged and went off in search of Gary.

Fliss looked at her friend. 'Are you sure there's nothing you want to talk about, Lisa?'

'I'm sure.' She sighed. 'Look, Fliss, I had a nightmare and I'm tired and I've got things to think about, so d'you think you could just leave me alone for a while, huh?'

'Sure.' Fliss felt hurt. 'I'll leave you alone. I'll stop talking to you altogether, if that's what you want.' She spun on her heel and hurried on down the drive.

CHAPTER TWELVE

'Right!' Mr Hepworth rubbed his hands together and beamed at Year Eight. 'It's just a week now since Mrs Evans and I sprang on you the task of producing a play for the Festival, and we thought this might be a good time for people to report back on how things are progressing. Not to us – we're here in an advisory capacity only – but to one another. Now – who'd like to kick us off?'

'I'd like to kick you off a cliff,' whispered one of the boys. His friend giggled.

Mr Hepworth glared at them. 'Did you speak, Roger?'

'No, Sir.'

'Then it was you, Michael. What did you say?'

'I – I said I'd like to kick us off, Sir.'

'Splendid – off you go, then.'

'Well, er – I'm a villager, Sir.'

'Yes?'

'And – my mum's nearly finished my outfit. She's made it out of sacking, and it's this raggedy old jacket thing with a belt and some really baggy trousers.'

'In other words, Michael, you'll be dressed much as usual.' Everybody laughed. 'And you, Roger – what are you up to?'

'I'm a Viking, Sir. I can sew a bit so I've done my own costume. Well – my mum helped a bit. And I've made this really wicked helmet, Sir, with wings on it.'

The teacher sighed. 'There's absolutely no evidence that the Vikings wore winged helmets, Roger. It's a fallacy.'

'No, it's a helmet, Sir, honest.'

'Yes, all right, Roger.' Mr Hepworth sounded tired. 'Sarah-Jane – you're the producer or director or whatever, aren't you?'

'Yes, Sir.'

'So how's it coming along?'

'Well – we thought about speaking parts, but in the end we decided to have a narrator because nobody knows how people spoke in those days.'

Mrs Evans nodded. 'Good idea, Sarah-Jane. Who's narrating?'

46

Andrew Roberts raised his hand. 'Me, Miss.'

Mrs Evans nodded. 'I can't say I'm surprised, Andrew. You've spent most of your time in this school narrating when you should have been listening. Go on, Sarah-Jane.'

'We've had a couple of rehearsals, Miss. Well – not really rehearsals. Trying things out, and it seems OK so far. We don't have people's costumes at school, and of course the worm's not ready, but—'

'It nearly is,' interrupted Trot. 'We've got everything. Now all we have to do is fit the skin and figure out a way to make it breathe fire.'

'Just a minute, David.' Mr Hepworth smiled. 'I know we want this worm to look as realistic as possible, but I think we're going to have to draw the line at fire-breathing.'

'Aw, Sir—?'

'No, David, I'm sorry. Anything you could devise would be highly dangerous. Just think what would happen if the worm caught fire with people inside it. It might be possible to fake smoke using dry ice or something, but there's to be no fire. Is that understood?'

'Yes, Sir.' Trot looked crestfallen.

'I thought he was here in an advisory capacity,' hissed Neil Atkinson.

The teacher looked up sharply. 'He is, and he has sharp ears, and his advice to you is to keep

47

comments of that sort to yourself. All right?'
'Yes, Sir.'

From the start of the session, Lisa had struggled
silently with herself. A part of her wanted to with-
draw from the play, or at least from the worm, while
another part – a dark, submerged part of herself whose
existence she hadn't even suspected a week ago –
urged her in excited whispers to say nothing: to hold
on to her place inside the worm and see where it might
lead her. And this had nothing to do with Trot. She
was fond of him, of course, but this was something
else; something altogether darker, more compelling.
And the dark part won. When twenty to three rolled
round and the teachers brought the session to an end,
she'd said nothing.

I'll do it, she cried inwardly, and a tingle ran
down her spine into her tummy-muscles. When Fliss
approached her gingerly at home-time she seemed her
old self, and they chatted as they dawdled up the
drive. Only Lisa knew she'd given in to something
dark and strong, and neither girl knew their paths
were set to diverge, or that when they came together
again it would be as enemies.

CHAPTER THIRTEEN

When Fliss got to Trot's at twenty-five past seven, Lisa was already there. Fliss's arms ached from carrying the curtain. She let it fall to the floor. Ellie-May and the two boys went down on their knees to look and feel. 'It's terrific, Fliss!' cried Gary.

Ellie-May lifted a fold, rubbed it against her cheek and let it slip through her fingers. 'Yeah, terrific. It's shiny, like it might be covered with slime or something, and the colour's exactly right. How long is it?'

They measured, and the curtain proved to be more than ten metres long. Lisa pulled a face. 'Pity. The real worm was twice as long.'

'Yes,' said Fliss, 'but remember it's very wide.

49

If we cut it in two lengthways and stitch the halves together we'll still have plenty of width and it'll be just the right length.'

'Who's going to do all this stitching?' asked Gary. 'I'm useless at sewing.'

'No problem,' Trot told him. 'My mum's volunteered to help. All we have to do is tack it more or less as we want it and she'll stitch it properly on the machine. Let's get started.'

It wasn't nearly as easy as it sounded. Their lengthways cut was a bit wavy and it took ages to fasten the two pieces together, even with Trot's big tacking stitch. Then they had to drape the slippery material over the worm's neck and body and mark where they'd attach the tapes which would be knotted under the worm's belly to fasten the skin to the frame. When that was done, they spread the fabric on the floor and took turns sewing on the tapes in such a way that they wouldn't be visible to an audience. It was twenty past nine by the time they'd finished, and they'd done nothing with the fifteen metres of material which would form the monster's tail. 'Leave that,' said Trot. 'My mum's clever. She'll figure out a way to cut and stitch so it tapers to a nice sharp tip.'

'Let's try it out,' suggested Gary. 'Fliss can tie the tapes, and the end can just trail for now like a peacock's tail. What d'you say?'

'I say yes!' cried Lisa, eyes shining.

'OK,' said Trot, 'only don't step on the tail or it'll rip off and all my brill tacking will go to waste.' He turned to Fliss. 'Will you do the tapes?'

Fliss shrugged. 'Sure, but don't be too long, OK? I was supposed to be home for half-nine.'

The four stood in line and lowered the frame over their heads while Fliss held the skin to stop it sliding off the hoops. Yells and laughs came from inside the worm as Fliss knelt, pulling down on the tapes and tying them. 'Hey, it's dark in here!' complained Ellie-May. 'I can't see where I'm going.'

I can see for all of us,' said Gary from the front. 'Put your hands on Trot's shoulders, Ellie-May, and go where he goes. Trot puts his on Lisa's and Lisa has hers on mine. Easy-peasy.'

It wasn't easy. Not at first, within the confines of the Trotter family's garage. Peering through the eye-holes on the worm's neck, Gary went off at a slow walk, twisting and turning to avoid walls, worktops and obstructions on the floor. The others followed as best they could, with frequent exclamations and much giggling. Fliss leaned against the workbench and watched. She wished they'd stop now so she could undo the tapes and go home, but they didn't.

At twenty to ten, Gary broke into a slow trot and the others followed suit. The worm danced sinuously through the darkening garage, its great

head swaying and bobbing. Now and then its reflector
eyes would catch light from somewhere and flash red.
Fliss was amazed at the dexterity of her friends; their
co-ordination. The way their dancing feet avoided the
great train of fabric they trailed, which slid, hissing,
across the dusty concrete. The ease with which they
seemed to have mastered the technique. Their shouts
of laughter grew louder as Gary increased his speed,
but there were no disasters – nobody stumbled.
Fliss watched as though mesmerized, and when she
remembered to look at her watch it was ten to ten.

'Hey!' Their exultant laughter drowned her voice.
'Hey, you guys. It's almost ten. I've got to go.'

Nobody heard. Gary shifted up another gear and
they came whooping in his wake, precisely, like a
well-drilled squad. Fliss moved over to the wall
switch and snapped on the lights. At once and in
unison the dancers broke into a rhythmic chant of
'Off, off, off!'

Fliss shook her head. 'No – it's ten o'clock.'

'You what?' cried Gary, and the others took it
up: 'You what, you what, you what?'

'I have to go.' She was close to tears.

'Go, go, go!'

'Lisa?' Surely her best friend would respond –
break step so that the dance could end in red-faced,
panting laughter?

'Lisa?' they mimicked, and her voice was among

them. 'Lisa, Lisa, Lisa, Lisa, Lisa—' The worm was coming at her now, eyes burning, jaws agape.

She turned and fled.

CHAPTER FOURTEEN

'Ceridwen, Ceridwen.' Mockery in their eyes, their voices.

'The worm. Terrific skin.'

'Triffic, triffic, triffic.'

'Tied with tapes.'

'Tape-worm, then. Heeee!'

'People inside, see?'

'Room for another though.'

'Room for one inside.'

'You, Ceridwen. Room for you.'

'I'm not Ceridwen!' she screamed. 'I'm Fliss.'

'Fliss!' they cried. 'Flass, Fluss, Floss.' Pressing in, crowding her so that she was forced to move out to where the worm danced. There it was. Its red, mad eyes and pinky, fang-crammed maw. It saw her and came slithering on a zig-zag path towards her. She tried to throw herself back, but they caught her and flung her forward again. The worm was close now. So close she could smell the putrid stench of its breath. Its slavering jaws gaped to engulf her. 'Room for one inside.' The voice was Gary's.

Fliss woke, damp and shaking. It was a long time before she slept again.

CHAPTER FIFTEEN

Tuesday morning. For the first time ever, Fliss didn't want to meet Lisa at the end of the road for the walk to school. She dawdled so long over breakfast that her mother started giving her funny looks. 'Fliss,' she said. 'Are you feeling all right?'

'I'm fine, Mum.' Pushing Coco Pops round her bowl.

'Then eat your breakfast, dear. It's almost twenty to nine. Lisa will go without you.'

That's the general idea, she thought, but didn't say. Her mother dropped toast on her plate. Fliss pushed aside the unfinished cereal and began to butter a slice as carefully as if she were painting a masterpiece. Her

56

mother sighed, cleared Dad's place and ran water into the sink.

Fliss knew her tactic had failed the moment she turned out of the driveway. The end of the road was about a hundred metres away and Lisa was there, waiting. It was almost ten to nine, for Pete's sake. They'd practically have to run to reach school on time, yet there she was. Fliss thought of ducking back into the driveway but if she did she'd certainly be late for school, and anyway Lisa had probably spotted her. With a grimace of resignation she walked towards the girl she'd regarded till lately as her best friend.

'Hi, Fliss. Why d'you leave in such a rush last night?' Lisa sounded genuinely concerned.

Fliss gazed at her. 'Are you kidding? After the way you all ignored me and mocked me and then came at me as though you meant to trample me into the floor? You'd have left in a rush too. Anyone would.'

'Would I heck!' Lisa's tone was scornful. 'It was a bit of fun, that's all.'

'Well, it wasn't fun for me, Lisa. It scared me, the way the four of you moved in that thing as though—'

'As though we'd been doing it all our lives,' finished Lisa. 'That's what you were going to say, isn't it?'

Fliss nodded. 'Something like that, yes.'

'And that's exactly how it felt, Fliss.' Lisa's eyes

shone. 'We couldn't put a foot wrong, any of us. I mean, you'd think— I expected we'd stumble and fumble around, you know? Knock things over, step on our own tail, fall down. Three of us couldn't even see, and yet we ended up running, Fliss. Running like one creature, not four. I can't describe the feeling except to say it was awesome. Sincerely awesome.'

'Yes, well, like I said, it was no fun for me.'

Lisa laughed. 'You shouldn't have joined if you can't take a joke, Fliss. And anyway, you'll get the last laugh, won't you?'

'How d'you mean?'

'You're Ceridwen, aren't you? Heroine-Saint of Elsworth? You get to vanquish the worm, remember?'

'Oh, yes. I see what you mean, but I still don't like the way you ganged up with the others against me last night, Lisa. You're supposed to be my friend.'

Lisa sighed. 'I am your friend, Fliss. Same as always, only you're not in the worm, see? You don't know what it's like 'cause you're not part of it, and that's bound to make a difference, right?'

Fliss shook her head. 'I don't see why. It's only a play when all's said and done.'

'Ah, but is it?'

'What d'you mean? Of course it is.'

'I dunno – maybe it is, maybe it isn't. Inside that worm last night it felt like something bigger, Fliss. Much bigger.'

'I don't know what you're talking about, Lisa. You've been talking crazy-talk ever since this play thing started and I wish you wouldn't. It scares me. I'll be glad when the Festival's over and the worm's gone for good.'

'If.'

'Huh?'

'If, not when. How do you know the worm'll go? It might win this time.'

'Don't be daft.'

Lisa shrugged. 'OK.' She looked at her watch. 'Two minutes to nine. Last one in school's a creepazoid.' She broke into a run and Fliss followed, wondering what old Hepworth would say when she told him she wasn't going to play Ceridwen.

CHAPTER SIXTEEN

The plate on the door said 'Deputy Head'. Fliss knocked. 'Come in.' She pushed open the door. Mr Hepworth smiled from the swivel chair behind his cluttered desk. 'Now then, Felicity, what can we do for you?'

'I don't want to be Ceridwen in the play, Sir.'

'Why ever not?'

'I don't really know, Sir. I mean, I know it sounds daft but I had this dream. This nightmare, about the worm. It scared me. And then last night—'

'What about last night?'

'Well, I don't want to get anybody into trouble, Sir, but something happened last night at David Trotter's and that scared me too.'

Mr Hepworth leaned forward across the desk. 'What sort of something, Felicity?'

'The worm, Sir. We finished the worm and they got inside it and—'

'Who? Who got inside it?'

'Ellie-May Sunderland, David Trotter, Gary Bazzard and Lisa Watmough, Sir. They're playing the worm.'

'I see. Go on.'

Fliss related the evening's events, including her flight from the garage. When she'd finished, the Deputy Head nodded. 'I can see how a thing like that might upset you, Felicity, but I'm not altogether surprised that it happened, considering who was in control of the worm.' He sighed. 'Whatever possessed Year Eight to put Gary Bazzard in the worm's head?'

'Well, he wanted to be the Viking Chief, Sir, but we'd decided to have a girl for that part, so Gary got the worm's head as a sort of consolation.'

'Well, it's Year Eight's production and we promised not to interfere, but I have to say that Mrs Evans and I probably would not have set our hearts on consoling Gary Bazzard, Felicity. The class gives him a leading role and he shows his gratitude by intimidating you with what sounds like a typical display of hooliganism. That's the sort of lad he is, I'm afraid.'

Fliss shook her head. 'It's not that that bothers

me, Sir. Gary's all mouth. I can cope with him any time. It's – other things that have happened. Things that have been said.'

Mr Hepworth shook his head. 'You're going to have to explain that, Felicity. You've lost me somewhere along the line.'

Fliss tried, but the things she had to say sounded ridiculous even to her, in the Deputy Head's office in broad daylight. The way they'd found everything they needed to make the worm. Lisa's remarks about fate. How the creature had turned out perfect without any striving on the part of its makers, and how easily the four children had learned to work it, as though they'd been doing it all their lives. And the change which seemed to have come over Lisa since she'd become involved. It was worrying stuff when you put it together but she spoke stumblingly and without conviction, presenting the teacher with a hopeless jumble of suppositions. When her voice tailed off in mid-sentence, he smiled.

'It's up to you, Felicity, but if you want my opinion it's this. Both you and Lisa Watmough have highly developed imaginations, and you've allowed them to run away with you a little. This, coupled with Gary Bazzard's typically idiotic antic, has given rise to needless anxiety on your part, the upshot of which is that you now wish to relinquish your part in the play.' He smiled again and shook his head. 'I don't think you should do that, Felicity. I feel you'd regret it later, when Year Eight's production turns out to be the highlight of the Festival. No. If I were you I'd be inclined to carry on. Put a bit of a curb on that imagination of yours, and remember that life is full of coincidences which may seem to add up to more than coincidence when you get a string of them together. And if I were producing this play, which I'm not, I'd stick Richard Varley in the worm's head and demote young Bazzard to understudy.' He arched his brow. 'All right, Felicity?'

Fliss nodded, looking into her lap. She wasn't convinced, not really. But she'd failed to convince the teacher so perhaps he was right. It did all seem a bit far-fetched now. Very far-fetched, in fact. She looked up. 'I'll try, Sir,' she said.

CHAPTER SEVENTEEN

She hadn't long to wait. While she'd been seeing the Deputy Head at morning break, Trot and Gary had collared Sarah-Jane and told her Trot's mum was finishing the skin, and that the worm would be complete by midday. Sarah-Jane persuaded them to go and fetch it in their lunch break so they could have a run-through with it. Then she went along to the staffroom and persuaded Mrs Evans to release Year Eight from English that afternoon for a rehearsal.

When Mrs Trotter's car pulled into the parking lot at ten past one that day, a crowd gathered to watch Trot and Gary unload their creation. It was in three separate pieces, but one of those pieces – the papier-mâché

head and neck – was impressive enough to draw
gasps and whistles from the watchers. 'Woweee!'
cried a first-year kid. 'Look at it – it's so real, like
they chopped the head off an actual dragon.'

'Yeah,' breathed another. 'And look at the eyes,
man. They stare at you, don't they? I reckon they
can see.'

A posse of kids trailed after the two boys as
they lugged their burden up the steps. They'd have
followed right into school if two prefects hadn't been
guarding the door. Trot and Gary were stowing the
worm in the Year Eight stockroom when Mrs Evans
came in. 'So this is it, eh?' She gazed at the head.
'Ugh!' She shivered. 'I wouldn't want to meet that on
a dark night, David. Did you do all this yourselves?'

'Yes, Miss.'

'Well, you've made a really good job of it, I'll
give you that. Papier-mâché isn't the easiest stuff to
work with, and those eyes are most effective. What
are they?'

'Car reflectors, Miss.'

'Car reflectors. Yes. Well – I can't wait to see
the creature in action.' She smiled. 'Well done.'

'Ta, Miss.'

'Ta?' Mrs Evans shot Gary a disapproving look.
'Surely you mean "Thank you, Miss"?'

'Oh – yeah. Thank you, Miss.'

'Hmmm.'

As Mrs Evans left the room, Mr Hepworth stuck his head round the door. 'What's this I hear about a monster?' His eyes fell on the head. 'Good heavens.' He came forward, stretching out a hand to touch its glossy skin. 'You've done a remarkable job here, lads. I don't wonder young Felicity had a nightmare.'

'Did she, Sir?' Gary's face was all sweet innocence.

The teacher looked at him. 'Yes, Gary Bazzard, she did, and if I hear of any more hooliganism on your part, you'll be out. Not only out of the worm but out of the play completely. I'd have thought by Year Eight you'd have grown out of that silly behaviour.'

'What silly behaviour, Sir? I don't know what you mean.'

'Yes you do, and it's got to stop. As of now. Understand?'

'Yessir.'

Mr Hepworth departed and they finished stowing the worm. As Trot closed the stockroom door, Gary spoke softly. 'There's a snitch in our midst, Trot. A tattletale.'

Trot nodded. 'Sounds like it. What we gonna do?'

'Oh, I dunno, Trot. I'll think of something.' Gary smiled. 'Something messy, I shouldn't wonder.'

CHAPTER EIGHTEEN

The hall was in use for PE, so they had to use Mrs Evans's room for the tryout. She supervised the stacking of chairs and tables along one wall to make floorspace, then disappeared in the direction of the staffroom with a pile of marking.

'Right.' Sarah-Jane perched herself on a window-ledge to do her producer bit. 'How are costumes coming along?'

'Mine's ready,' said Gemma, 'but it's at home.'

Fliss nodded. 'Mine too. Nobody said we were rehearsing today.'

'I know,' grinned Sarah-Jane. 'It was a spur of the moment decision. I couldn't wait to see the worm in action. Has anybody brought their costume?'

Nobody had, but it didn't really matter. The only costume anybody was interested in at the moment lay in three pieces in the Year Eight stockroom. Trot and Gary carted it out and there was no shortage of volunteers to help Ellie-May, Lisa and the two boys into it. When the last tape was tied, Gary led his team on a trial circuit of the classroom under the admiring gaze of their classmates. Mrs Trotter had stuffed and sewn the long tail beautifully. It was rounded and tapered and flexible and it looped and snaked across the floor as the monster circled.

'OK,' said Sarah-Jane, when the worm had done three circuits. 'That's beautiful, but I got us off English and we're supposed to be working.'

'Let's do the bits where the worm seizes villagers and drags them off,' suggested Keith.

There was a general cry of 'Yeah!' and Sarah-Jane nodded, pointing. 'That's the village, over in that corner. Get over there if you're a villager.'

'Which bit's Norway?' demanded Barry Tune. Sarah-Jane looked at him. 'What d'you mean, which bit's Norway? What's Norway got to do with it?'

'That's where the Vikings were when the worm was eating peopleburgers,' said Barry. 'So that's where us Vikings should stand.'

'Don't be daft,' snapped Sarah-Jane. 'The Vikings aren't in this bit. They can stand round the walls and watch.'

68

'Boring,' muttered Barry. 'If there's one thing a Viking hates, it's being bored.' Some of the other Vikings muttered their agreement. Sarah-Jane ignored them. Meanwhile the villagers had crammed themselves into their corner and were arguing over who should be the first victim, while the worm glared balefully at them through its mad red eyes.

After some pushing and shoving, Tara Matejak was thrust forward by Michael who cried, 'Here's your starter, worm.'

'Just a minute!' Gary, who was moving towards the girl, stopped at the sound of Sarah-Jane's voice. Sarah-Jane glared at Michael. 'Is that what you intend saying on the day, Michael Tostevin?'

The boy grinned. ''Course not.'

'Then don't say it in rehearsal, OK?'

Michael shrugged. 'OK, Miss. Sorry, Miss.' Some of his friends tittered.

Sarah-Jane sighed. 'OK, worm – carry on.'

When it came to it, the business of seizing and dragging off proved far more difficult than anyone had envisaged. The jaws of the monster were not a moving part. They were set permanently agape and could seize nothing, so that Tara had to co-operate in her own abduction, thrusting her hand into a corner of the worm's mouth and walking beside it in such a way as to suggest that she was being dragged by the arm. It wasn't completely successful, and Trot undertook to devise a way of enabling the beast to grab its living meals more convincingly in future.

Fliss observed all of this with apprehension, praying that time would run out before Sarah-Jane decided enough villagers had perished and called upon Ceridwen to confront the worm. She'd promised Mr Hepworth she'd try, but her classmates' dexterity inside that awful disguise disturbed her even here, and she was far from happy. It must have been her lucky day, because the buzzer went as the beast prepared to bear away its sixth victim.

'OK.' Sarah-Jane slid down from her perch. 'Wrap it up, everybody.' She smiled. 'That wasn't bad, but I want all costumes in school tomorrow.' She turned to Trot, who was struggling out of his disguise like a moth from a chrysalis. 'Don't forget, Trot – the worm needs to be able to grab its prey.'

Trot nodded. 'I'll think of something.'

Sarah-Jane turned to speak to Fliss, and was mildly irritated to find she was no longer in the room.

CHAPTER NINETEEN

She was passing the Deputy Head's office on her way out when the door opened. 'Ah, Felicity, come in a minute, will you?' Mr Hepworth stepped to one side and she went in. He closed the door and stood with his back to it. 'Now – how did the rehearsal go?'

'All right, Sir.'

'No trouble from our friend Mr Bazzard?'

'No, Sir.'

'Good. I had a word with him and it seems to have worked. So, are you feeling a bit happier about things now, Felicity? We wouldn't want to lose your talents, you know.'

Happier? Fliss would have laughed out loud if she'd dared. Mr Hepworth had had a word with

Gary, which meant Gary knew she'd complained. He'd have her marked down as a sneak. He'd tell the others. Her name would be mud.

'I – dunno. Sir. We didn't get to my part. I'll try.'

'Good girl.' He opened the door. 'Off you go, then. And let me know if you have any more hassle.'

'Yessir. G'night, Sir.'

'Goodbye, Felicity.'

'Let me know if you have any more hassle.' That's a laugh for a start, she thought. I can tell you now there'll be hassle, but there's no way I'm gonna let you know. No way.

The drive was thick with pupils going home. Fliss dodged between them, hurrying, looking for Lisa. Lisa knows how these things happen, she thought. She'll understand. I'll tell her I didn't mean to get anybody into trouble. It just came out.

She was through the gateway and well along the road before she spotted her friend. Lisa was walking with Ellie-May Sunderland. They were dawdling, deep in conversation. Fliss put on a spurt and caught up. 'Hi, Lisa, Ellie-May.'

The two girls regarded her coldly. 'What do you want?' asked Lisa.

'I've got something to tell you.'

'We're talking. See you tomorrow, OK?'

'What's up – what have I done?'

'You know.'

'No I don't.'

'You split on us to old Hepworth.'

'No I didn't. Not on purpose. That's what I wanted to talk to you about.'

'We're not interested in excuses, Fliss. You split on us. That's all that matters.'

'Yes, but—'

'No buts.'

'Are you out tonight, then? We could—'

'No. We're busy tonight, working on the worm.'

'I'll come to Trot's then, shall I?'

Lisa laughed. 'I wouldn't if I were you, Fliss.'

'What d'you mean?'

'What d'you think I mean? Gary's after you, dummy. He'd love you to show up at Trot's. You'd come on foot and leave in an ambulance.'

'But what about you, Lisa? You're not Gary. You don't have to do everything he does. We're friends, aren't we?'

'No, Fliss, we're not, since you ask. Why don't you get lost and leave us in peace?'

'I—' Fliss realized with horror that she was about to cry. Biting her lip she turned away and crossed the road, half-blind with tears. There was an entry – a narrow walkway between two buildings which led on to waste ground. She turned into it, away from the stream of chattering kids, and when she was alone, she wept.

That night, Fliss dreamed again. She'd grown since
her bridesmaid day. The long white dress no longer
covered her ankles, so Mum had let down the hem to
lengthen it. Now she wanted Fliss to try it on, but the
alteration had transformed the dress. Mum couldn't
see it – she was holding the thing out for her to slip
into – but it wasn't a dress any more. It was a—

'A shroud!' She was screaming, shaking her head.
'Can't you see, Mum? It's a shroud.'

'Don't be silly, dear. Come – try it on.' Mum
advanced on her, smiling.

'No.' Backing away, hands out to ward off
the loathsome garment. Bitter tang of tears in her
mouth. Backing towards the door, which opened.
Mr Hepworth came in, smiling. 'Try, Fliss,' he
crooned. 'Try it on. It is like a shroud, but life
is full of coincidences.'

'No, I don't want to. Leave me alone.'

'Typically idiotic antic.'

They rushed, seized her. She struggled, but the
Deputy Head was holding her from behind and Mum
had the cold fabric over her head. It clung, reeking of
sodden clay, smothering her. She jerked herself this
way and that. Couldn't breathe. Dark rising. Can't
breathe can't breathe can't breathe —

She woke with her face pressed in the pillow
and the bedclothes on the floor.

CHAPTER TWENTY

Ronnie Millhouse was the town drunk. Everybody knew him by sight – he was what is known as a 'character' – but nobody knew the trouble he'd seen. Like all drunks he'd once had an ordinary life, but then the trouble had struck and he'd taken to the lotion in a big way. Now he spent his days on the street, cadging ten and twenty pence pieces from passers-by. 'Have you got any spare change?' he'd ask. 'A few pence for a cup of tea?' People either brushed past him looking angry, or fished in their pockets looking embarrassed, and most days there were enough of the latter sort to provide poor Ronnie with the price of several cups of tea. He didn't waste it on tea, of course. Ronnie's refreshment usually came in a fat brown bottle with a picture of a woodpecker on it. At night, when the wind blew chill and the stream of passers-by dried to a trickle, Ronnie would make his way to the derelict bandstand in the park, where he had a cardboard box for an hotel and a drift of old newspapers for his bed.

At eleven-thirty that Tuesday night, while Fliss lay dreaming, Ronnie was shuffling unsteadily along the footpath which led to the bandstand. A fine drizzle was falling. On his left was the kiddies' playground where the swings hung motionless on dripping chains and the slide gleamed wetly in the light from a distant streetlamp. To his right, the ground fell away in a long slope, thickly planted with trees and shrubs. At the foot of this slope, hidden even in daylight by the trees, was a stretch of level grassland on which, from time to time in the summer months, funfairs and circuses would pitch their camps. Now, as he headed for his bed at the end of a better-than-average day, Ronnie thought he heard voices on the slope. Now Ronnie was a cautious man even when drunk, and he knew there was a better-than-even chance that anybody you'd meet in a public park late at night would be up to no good, so he swerved off the path and pressed himself up against the wet trunk of a thickish tree to see who might appear.

There was a scraping, crackling sound like something big in the shrubbery. Whatever it was, was coming up the slope pretty fast. Ronnie pressed himself more closely to his tree and peeped round, and it was then he saw the dragon. He screwed up his eyes and shook his head and looked again but it was still there, coming off the slope on to the foot-path. Its teeth gleamed white and its eyes blazed red. He couldn't make out its colour in the dark, but as it crossed the path and headed for the playground he saw that it was incredibly long. He stood absolutely still and held his breath as the monster's whiplike tail hissed across the tarmac. He hugged his tree while the great shape crossed the playground, nor did he stir for some time after darkness swallowed the beast and all was quiet.

When Ronnie finally let go of the tree and resumed his journey it was twenty minutes before midnight. Half a mile away, Fliss had just woken from her nightmare. It was still drizzling.

Ronnie reached the bandstand and got into bed. He lay on the dusty boards and thought about the dragon. For a while he told himself he'd report what he'd seen. He'd tell the police or the local paper. His fuddled brain created a fantasy in which for once, people were interested in him. A fantasy in which he was somebody because of what he had seen.

It soon faded though. He'd had a good day. A two-bottle day. Who needs fame when there are bottles waiting to be drained? And what's a dragon, compared to some of the creatures Ronnie

Millhouse had seen? Pink lizards. That kangaroo in pinstripe suit and bowler who'd tipped him a fiver. The bright green ants who sometimes ate his hands. No. He'd not tell. Why should he? Waste of time.

Drizzle fell endlessly. Wind lifted a corner of his paper blanket. Ronnie Millhouse slept.

CHAPTER TWENTY-ONE

'Morning, Mum, Dad.'

Lisa sat down, reached for the packet and sprinkled cornflakes in her bowl. She'd overslept. Dad was halfway down his second cup of coffee and Mum had had to call her twice. She avoided their eyes, hoping they'd say nothing, but it was a forlorn hope.

'Tired, are we?' her father enquired.

'She ought to be,' said her mother, 'coming in at midnight, bold as brass, saying she's been busy. I'll give her busy if it ever happens again.'

Her husband nodded. 'Where was she, that's what I'd like to know.'

Lisa sighed. She hated it when her parents discussed her as if she wasn't there and besides, they'd been over all this last night. 'I told you,' she mumbled. 'I was at David Trotter's, working on the worm.'

'Till midnight?'

'Yes, Mum. It was a big job.'

'It must have been. I'm surprised at the Trotters, letting kids your age stay out till that time. Didn't they realize we'd worry?'

Lisa shook her head. 'They were out, Mum. I told you.'

'We know what you told us, young woman,' rapped her father, 'and now I'll tell *you* something. If anything of this sort happens again I'll be along to school to see Mr Hepworth, and we'll have you out of that play. I'm not having a daughter of mine staying out all night at thirteen years of age, no matter how busy she is. Do you understand?'

'Yes, Dad.'

'Well, I certainly hope so. And I hope you can attend to your lessons today without falling asleep at your desk.'

It was nearly ten to nine when Lisa finally got out of the house. It was still raining, and she wasn't surprised to find no Fliss waiting at the end of the road. She wasn't surprised, and she didn't care. She didn't want to talk to Fliss. It would be no use talking to her. Fliss didn't know. She hadn't been there. You had to have been there to know how it felt, running through the dark. The dark in the park. She smiled briefly at the unintentional rhyme. The park after dark, where you'd hardly dare venture in ordinary circumstances because of the hooligans and the glue-sniffers and the funny men Mum was always on about. You'd stay away if you'd any sense, unless you were part of the worm.

Part of the worm! She laughed out loud, remembering. What a fantastic feeling, running through the dark, fearless because you are part of the most fearsome thing in the park. Fearless because nothing exists which can harm you. There is nothing which wouldn't run screaming and blubbering at your approach. Hooligans, glue-sniffers, funny men. All fleeing, fleeing before – before ME! Lisa's exultant laugh turned a few heads among early shoppers as she ran with her head thrown back and her hair flying, to recapture a scrap of last night's fierce, narcotic joy.

CHAPTER TWENTY-TWO

'"The time – a little over one thousand years ago. The place – Elsworth, then a mere village, set in the midst of —"'

'Can't hear him!'

'Speak up, you mumbling creepazoid!'

'OK, you lot, you've made your point.' Sarah-Jane looked across the field to where Andrew Roberts stood between the goal-posts, clutching his script. 'You'll have to project a bit, Andrew. Remember we're outside and there's a bit of a breeze.'

The narrator nodded. 'Shall I start again, then?'

'Please.'

It was lunchtime. Year Eight, resplendent in full costume, were rehearsing on the school playing field before a packed audience. It had stopped raining only an hour before and the grass was wet, but Sarah-Jane had been determined and so here they were, Vikings and villagers, saint and serpent, in full regalia, hoping to get through the whole thing before the bell.

' "The time – a little over one thousand years ago. The place – Elsworth, then a mere village, set in the midst of misty fenland. Elsworth, a once quiet village where terror now reigns, for the nearby fen has become the dwelling-place of a monster – a monster known to every terrified inhabitant as THE WORM." '

Behind the narrator and his goal-posts, the field fell away in a steep grassy bank to the stone wall which at this side marked the boundary of school property. After rain, the strip of land between the foot of the bank and the wall became waterlogged, forming a moat of brown water and sticky mud. As Andrew spoke the worm appeared, lurching up the bank to the cheers and whistles of the watching multitude before trotting the length of the field on eight muddy feet to assault the goal-mouth at that end, which was crammed with villagers. This time, Joanne O'Connor was selected as the creature's first victim and pushed out towards the penalty spot. The worm ran at the girl as if it meant to boot her into the back of the net, but at the last moment Gary reached out

83

and grabbed her. The crowd roared, drowning Joanne's half-genuine scream as she was hustled over the halfway line with her feet off the ground.

'Right!' Sarah-Jane flapped a hand at the worm. 'You can put her down now, Gary – we get the idea. Brilliant solution by the way, but time's short. Can we go to Ceridwen please?'

Fliss, who'd been standing on the touchline trying to keep the hem of her dress clear of the mud, felt her heart kick. She'd known this moment would come, but had expected Sarah-Jane to allow the worm a few more victims before calling on her. She caught her bottom lip between her teeth and stepped forward, holding a plastic sword, hoping nobody would notice her nervousness.

'Gary.' Sarah-Jane gestured towards the banking from which the worm had made its entrance. 'Out of sight, please. Fliss – you walk out of the village while Andrew's doing his next bit and stand on the halfway line. When the worm rushes you, raise your sword as if you're going to slash at its neck. Gary.' The worm paused, seeming to glare at the director with its mad eyes. 'You shoot out your arms to grab her like you did with Joanne, but as soon as your hand brushes the dress you back off, looking submissive. Can you do that?' The worm made no reply, but turned and loped off towards the banking. When it was out of sight, Sarah-Jane nodded to the narrator, who began to speak. Fliss swallowed hard and set off for the middle of the field.

' "—armed only with a short sword and her faith, stood directly in its path." ' Andrew stopped speaking. The worm topped the rise to the cheers of the spectators and came trotting towards Fliss, its great head swaying from side to side. Fliss swallowed again, gripped her sword tightly and lifted it above her head.

That's when it all went wrong. Gary's arms appeared, but he didn't brush the dress. Instead, he grabbed her on the run and turned, and Fliss found herself being carried swiftly back the way the worm had come. She kicked and shouted and laid about her with the sword but it was no use. Gary's hold was like the hug of a bear. Ignoring Sarah-Jane's cries, he carried Fliss to the top of the slope, hissed, 'Tattletale!' through the eye-holes and flung her down. Helpless, she half rolled, half skidded down the banking and into the moat. The crowd, believing this to be part of the show, cheered themselves hoarse as she lay winded, feeling the spread of clammy wetness which would turn the white dress brown.

CHAPTER TWENTY-THREE

The show had to go on, and the Vikings were making their first raid as Fliss picked herself up and ran sobbing to the girls' changing-room. She pulled off the sodden dress and held it up. It was so obviously ruined that she flung it to the floor and flopped down on a bench, weeping. Her first thought was, Right – that's it. I'm out. As soon as I get cleaned up, I'm off to Hepworth to tell him I'm not doing it. She stripped for the shower, and as she stood under the warm torrent it occurred to her that Gary and the others might actually be trying to get rid of her. They want me to quit, she told herself. That's why Gary does rotten things to me while Lisa and the others ignore me. They want me out and Samantha in.

She didn't know how she knew this but it felt right, and it brought about a change of mind – a fierce determination. No, she thought, turning off the shower and rubbing herself with a scratchy towel, they're not going to force me out if that's their little game, because I won't go. I'll hang on. I won't even mention this rotten trick to Mr Hepworth or Mrs Evans. I won't tell Mum either. I'll say it was an accident. I slipped and fell down the banking. Mum'll know a way to save my dress. Next rehearsal, Gary Bazzard and his friends are going to find me there as if nothing's happened. And the one after that, and the one after that – right up to the great day itself. And if they don't like it, they can go take a running jump.

While Fliss was undergoing her change of mind in the shower, Gary, Trot, Lisa and Ellie-May were stowing the dismantled worm in the Year Eight stockroom. Their mood was subdued as they awaited the consequence of their leader's vicious act. 'You're an idiot, Gary,' said Ellie-May. 'I bet she's in with old Hepworth right now, laying it on. We'll all be out, you see if we're not.'

Gary shrugged. 'She asked for it, and anyway, I couldn't help it. Something came over me.'

Lisa shot him a venomous glance. 'Something came over you? We're gonna lose the best kick any of us ever had, and all you can say is something came over you?'

'I don't think she'll tell,' said Trot.

Gary sneered. 'You're joking.'

'No I'm not. I know Fliss Morgan. She's got a stubborn streak. You and Lisa have been making it pretty obvious we don't want her around. I reckon she'll stick, just to spite us.'

Ellie-May sighed. 'I hope you're right, Trot. I want to do the park again like last night. It was the most fantastic feeling I've ever had.'

'Well that's what I mean!' cried Gary. 'That feeling. It comes over you and you can't help what you do. It's – awesome.'

'Yeah!' Trot smiled dreamily. 'Maybe next time we'll run into somebody – somebody we can scare.'

'If there *is* a next time,' muttered Lisa.

The four spent the afternoon in a state of suspense, but nothing happened. Fliss avoided their glances in class and kept well away from them at break, but nobody was summoned to the Deputy Head's office and Mrs Evans gave no sign she was aware of anything amiss. The only sticky moment came at home-time, when Mrs Evans found Lisa carrying part of the worm through the girls' cloakroom.

'Where are you going with that, Lisa?'

'Taking it home, Miss.'

'What on earth for?'

Lisa's brain raced. 'Er – safety, Miss.'

'Safety? What d'you mean, safety?'

'Schools get broken into, Miss. We wouldn't want vandals smashing up our worm.'

'I see. So you intend carting the whole thing backwards and forwards every time there's a rehearsal?'

'Yes, Miss.'

'Hmmm. Well, rather you than me, Lisa, that's all I can say. Off you go.' She went out on to the step and watched for a moment as Lisa joined Gary and the other two in the yard and the four of them went off up the driveway with their burdens. She and the Deputy Head had undertaken not to interfere in the play – it was to be an independent Year Eight effort, and if the children had decided to keep some of their props at home, so be it. Lisa's reasoning seemed decidedly odd, but then a lot of the things children do seem odd to adults, and Mrs Evans supposed there could be no harm in what they were doing.

Which just goes to show how wrong you can be.

CHAPTER TWENTY-FOUR

Hughie Ackroyd hated kids. Until his retirement four years earlier he'd been a crossing keeper on the railway, and it seemed to him that he'd spent half his life chasing kids off the line and the other half making them stay off. The only thing he'd liked about his job was the bit of garden which went with the keeper's cottage. He'd kept that garden so beautiful that travellers in passing trains used to go 'Ooh!' and 'Aah!' as they whizzed by, and some of them would look back with their faces pressed against the window till Hughie's crossing was out of sight.

Now that he was a pensioner, old Hughie didn't have the garden any more. He and his wife lived in an old folks' bungalow. The grass outside was mown by the council, which also sent young men to tend the flowerbeds. Bored out of his skull, Hughie had taken an allotment on a nearby block and started growing his own vegetables. He'd turned out to be as good with vegetables as he used to be with flowers, and his leeks sometimes won prizes at local shows, which made him happy.

What didn't make him happy was this. One of the plots on the block was derelict. It had been derelict for many years and had become a jungle of couch-grass, weeds and brambles. This abandoned plot happened to be right next to Hughie's immaculate one, and in one corner of it stood a dilapidated greenhouse. This greenhouse had an old iron stove inside, and a bunch of kids sometimes showed up on wet weekends to light this stove and mess around in the greenhouse. They weren't doing any harm, except that occasionally, when there was nobody about, they'd pop on to somebody's plot and help themselves to the odd raspberry or handful of currants. They were trespassing though, and anyway Hughie hated kids. If they turned up when he was on his plot he'd shout over the rickety fence which separated his garden from the jungle, shaking whatever implement he happened to be holding, telling them they were trespassing and threatening them with the police. They'd gaze at him sullenly for a while then slink off through the rain, calling him rude names under their breath. This had been going on for at least two years, and the hatred he felt for them was matched by their dislike of him.

One of these kids was Gary Bazzard. Another was David Trotter. The rest were friends who attended a different school and went round with Gary and Trot at weekends and in the holidays.

Old Hughie's miserable face floated into Trot's mind that Wednesday evening when he, Gary, Lisa and Ellie-May were hanging around Trot's garden gate. Three weeks ago the girls wouldn't have been seen dead with the boys outside school hours, but lately the four had found themselves drawn to one another by an attraction each avoided thinking about, though they knew it had something to do with the worm. Mrs Trotter, watching them through her front window, told herself that if her son had started taking an interest in girls it was probably that Gary's fault, and decided to mention it to her husband.

'What we gonna do?' said Ellie-May.

Gary grinned. 'What d'you think?'

'The park, of course.' This from Lisa.

'No.' Trot shook his head. 'I've got a better idea.'

They all looked at him. 'What?'

'Old Ackroyd.'

Lisa frowned. 'Who's he?'

Trot explained. 'He practically lives on that allotment. He'll be there till it's too dark to see his stupid lettuces or whatever.'

'So?' Ellie-May looked quizzical.

'So we take the worm over to the allotments, get into it and spook the living daylights out of him. What d'you reckon?'

'I dunno.' Lisa pulled a face. 'He's old, you said. He might have a heart attack or something.'

'Will he heck! If he'd a bad heart, he wouldn't be able to dig that massive allotment, would he?'

Gary shook his head. 'He'd be at home all the time, watching telly and popping pills. I say let's do it.'

So they did.

CHAPTER TWENTY-FIVE

'Yes, Sir?' The young constable looked across the counter at the elderly man in grubby overalls. He couldn't see the man's boots, but he could see the muddy tracks they'd left on the gleaming lino tiles and they irritated him. There's a doormat, he felt like saying, so why don't you use it? He wanted to say that, but instead he said, 'Yes, Sir?'

Hughie Ackroyd glared. 'I want to report an act of vandalism.'

'What sort of vandalism, Sir?'

'Mindless vandalism, of course. The sort you get because bobbies don't walk the streets any more.'

'And where did this – vandalism occur, Sir? Were you a witness?'

'Of course I was a witness. It was my allotment, wasn't it?'

'I don't know, Sir.' The constable reached out, slid a thick notepad towards himself and fished in his pocket for a ballpoint. 'I think we'd better start at the beginning. Can I have your name, Sir?'

'Hugh Ackroyd.'

The constable wrote on the pad. 'Address?'

The man sighed. 'Twenty-two, Alma Terrace. Look – do we have to go through all this? By the time you've finished fossicking about, that dragon'll have vanished without trace.'

The constable looked up. 'Dragon, Sir?'

'That's what I said.'

'You want to report an act of vandalism by a dragon?'

'Yes. Well – it wasn't a real dragon, of course. It was kids dressed up.'

'Kids dressed up.' The policeman put down his pen. 'How many kids were there, Sir?'

'I dunno, do I? They were in this dragon thing. I were packing up for the night – hoeing my last row of spring onions – and this contraption comes running through the gate. It – they – trampled all over my beds, pushed my incinerator over and ran off laughing.'

'I see. At about what time was this, Sir?'

'What's that got to do with it?'

'It's procedure, Sir.'

'It's a waste of flippin' time, that's what it is. I might have known there'd be no point coming here. You're all too busy cruising about in your luxury limousines these days, talking into them poncey radios, so why don't you just forget it, eh? Pretend I never came in. I'll take care of this – my way.' He spun on one mud-caked heel and made for the door.

'I wouldn't advise—' The constable broke off as Hughie Ackroyd slammed out. 'Watch out for those dragons, Sir,' he murmured to the still-quivering door.

CHAPTER TWENTY-SIX

As Hughie Ackroyd was tracking mud into the police station, Trot was doing the same to the kitchen at home. His mother shrieked as he clomped across the floor. 'Look at the state of your shoes, David. Take them off at once and leave them on the mat.'

Trot turned with a sigh. 'Yes, Mum.'

'Wherever have you been to get them in that state?'

'Oh – around. You know.' Squatting by the doormat, fiddling with his laces. 'The park, mostly.'

'You must have been on the flowerbeds to get so filthy.'

'Maybe. We didn't mean to.'

'No. Anyway, your dad and I would like a word with you.'

'A word?' Trot's heart lurched. 'What about?' Surely old Ackroyd hasn't been here, he thought. He couldn't possibly know it was me.

'About you,' said his mother unhelpfully. 'Your dad's in the front room.'

Trot left his trainers on the mat and trailed after his mother. His father smiled up at him from an easy chair. 'Hello, son.'

Oh-oh. Trot returned the smile. Something's up. 'Hi, Dad.'

'Sit down a minute, David.' His father indicated the other chair. Trot sank into it, watching his parents' faces. They didn't look mad or anything. His mother sat down on the sofa.

'So, how're things going, son?'

Trot pulled a face. 'OK, I guess.' He couldn't remember the last time his father had asked him how things were going. There probably hadn't been a last time, so what was all this about?

'Good, good. The play?'

'Fine.'

'Your friend – Gary, is it?'

'He's fine too, Dad.'

'Good. I expect he's got a girlfriend, eh – good-looking lad like him.'

The way his father chuckled as he said this switched on a little light in Trot's head. Ah, he thought. So that's what all this is about. Girlfriends.

'Er – no.' He shook his head. 'Not that I know of.'

'Oh.' His father shrugged. 'It's just that your mother and I seem to have seen quite a lot of Lisa Watmough and the Sunderland girl just lately, and we wondered —'

'They're in the worm, Dad. We have to practise, y'know?'

'Oh yes, of course. So you're not particularly interested in either of them, then?'

Trot shook his head. 'No way. Ellie-May's a droop and that Lisa's got a face like the back end of a motorway pile-up.'

'David!' his mother frowned. 'That's not very nice, is it?'

'What – Lisa's phizog?'

'No – you know perfectly well what I mean. Talking like that. Lisa Watmough's quite a pretty girl. I was at school with her mother and she was pretty too.'

'Good.' He looked from parent to parent. 'Is that it, then? Can we have the telly on now?'

His father looked at him. 'Thirteen's a difficult age, son. You know you can always talk to me and your mum if anything's worrying you, don't you?'

'Sure I do, Dad. Nothing's worrying me, honestly.' Quite the reverse, he thought, recalling the expression

99

on old Ackroyd's face as he watched the worm mess up his stupid garden. Everything's fine. And it's going to get a whole lot finer.

'Good.' His father gripped the arms of his chair and levered himself upright. 'There's a film on Channel Four you might enjoy. I think I'll stroll down to the club for half an hour.'

When her husband had left the room, Mrs Trotter looked across at her son. 'Are you absolutely sure you're not fretting about anything, David?'

Trot grinned. 'Absolutely, Mum. There's nothing I can't handle. Nothing in the world.' As he said this, something occurred to him which wiped the grin off his face and caused his heart to kick. How is it, he wondered, that I saw the look on Ackroyd's face when only Gary has eye-holes?

CHAPTER TWENTY-SEVEN

Fliss's mum left the dress to soak over Wednesday night in a strong detergent, and when she lifted it out of the bowl next morning and held it up to the light, the stains seemed to have gone. 'We shan't know for certain till it's dry,' she cautioned, but Fliss smiled tightly and said, 'It'll be fine.'

Lisa wasn't anywhere in sight when she got to the end of the road, but when she was halfway to school she heard someone call her name. She turned. Vicky Holmes was hurrying to catch her up. 'Hi, Fliss,' she smiled, falling into step. 'I – I just wanted to say I think it's rotten what they did to you yesterday. That lovely dress.'

Fliss nodded. 'Thanks, Vicky. My mum washed it. It's going to be OK.'

'Yes, but still.'

'I know. Gary Bazzard's a pain. He's always been a pain, but he seems to have got a lot worse since we've been doing this play. The others have too. I think they're trying to get rid of me.'

'Rid of you – how d'you mean?' Vicky looked horrified.

Fliss grinned. 'I don't mean murder, Vicky. I mean they want me out of the play.'

'Why?'

'Dunno. I don't think they know either.'

Vicky looked at her. 'That's a funny thing to say.'

'Yes I know, but it's true. It's like something's gotten hold of them since they've had that costume. Look at Lisa Watmough – she was my best friend.'

Vicky nodded. 'I've noticed.' She laid a hand on Fliss's arm. 'I'm your friend, Fliss.'

Fliss smiled. 'I know, and I'm glad. I mean it.'

That afternoon there was a long rehearsal in the double-games period. Everybody was in costume except Fliss, who felt a wally in skirt and jumper, waving her plastic sword. She was apprehensive too but she didn't let it show, and when Gary reached for her she hissed, 'You dump me down that bank again and I swear I'll smash your stupid costume once and for all. You wouldn't like that, would you?' No reply came from inside the worm, but when Gary's fingers touched her sleeve the creature shrank back in a most convincing way.

'Begone, foul fiend!' cried Fliss, pointing her sword towards an imaginary fen. 'I command you – in God's name begone, and come this way no more.' Very quietly, through lips which scarcely moved she added, 'You don't get rid of me that easily, Bazzard.'

The monster slunk away.

CHAPTER TWENTY-EIGHT

Saturday morning, Fliss left the house at eight forty-five. It was day one of the Festival, and walking home together Friday afternoon she and Vicky had arranged to meet in Butterfield's diner to drink Coke and watch the procession with which the Festival was to open. Somebody at school had suggested putting the worm in the procession but Gary had dismissed the idea. He said it would spoil the surprise.

Elsworth was a small town and Butterfield's was its only supermarket. The diner which was tacked on its side was a favourite meeting place for kids. As she turned on to the road which dropped down into town, Fliss was thinking about the bridesmaid dress. Dry and ironed, it bore an indistinct mark where the edge of the stain had been, but this mark was so faint you'd have to know it was there before you'd see it. It certainly wasn't going to stop her wearing it for the play next weekend, so Gary Bazzard's dirty trick didn't matter any more. And, she told herself, since I've found a brand-new friend, Lisa Watmough doesn't matter either.

It was five past nine when Fliss reached Butterfield's, and Vicky was already there. She'd bagged a table by the window so they could watch the parade in luxury and pull faces at any boys who might go by. She had a can already, so Fliss got a Coke from the cabinet and paid at the counter before sliding in beside her.

'Hi, Vicky. Been here long?'

Vicky shook her head. 'Three, four minutes. Grant Cooper and Michael Tostevin just went by. They've gone to McDonald's.'

'How d'you know?'

'They mouthed it through the window. Probably hoped we'd join them.'

'No chance.'

'What'll we do after the procession?'

Fliss shrugged. 'Whatever you like, as long as it doesn't involve Grant and Michael. I see enough of them at school.'

They lingered over their drinks, turning and giggling when a knot of older boys looked in the window.

One of them was tall and lean, with thick black hair and a cheeky grin, and Fliss wished he'd come in and whisk her away to somewhere romantic, but he only stretched his mouth with his forefingers till it looked like a letterbox and wiggled his tongue at them. As he was doing this, the Mayor's limousine came cruising by at the head of the procession and the boy moved away, looking abashed. The girls' vantage point turned out not to be so great after all, because their view was partly blocked as shoppers lined the pavement to watch the floats. As soon as the last float had passed, Fliss and Vicky slurped up the dregs of their Cokes and went outside.

They strolled through the town. The spectators were dispersing, leaving crisp packets and bits of torn streamer on the ground. When they came to where the Odeon used to be, there was just a gap with bits of smashed masonry and the marks of heavy tyres. They stood for a while gazing at the gap, and Fliss told Vicky about the demolition man and the fabric he'd given to Lisa and herself.

They walked on, through the shopping centre and into the square. The parish church – St Ceridwen's – overlooked the square, and as the girls approached they saw that somebody had stuck a colourful poster on the notice board. They stopped to read it.

ONE THOUSAND YEARS IN ELSWORTH

it began. 'What a thought,' groaned Fliss. 'One Saturday morning's bad enough.'

'Yes, but look,' cried Vicky. 'It mentions our play.'

'Where?'

The poster listed a whole lot of things which would happen during the coming week. Fliss's eyes slid down the list. There was the procession they'd just watched, with a prize for the best float; a Festival Queen, whom they'd glimpsed enthroned on the back of a lorry; a knockout quiz competition; a prize for the most original shop window display, and much else besides. At the foot of the list, in brilliant green, was this:

SATURDAY MAY 1ST. ON THE FESTIVAL FIELD. A THRILLING RE-ENACTMENT BY CHILDREN OF BOTTOMTOP MIDDLE SCHOOL OF SAINT CERIDWEN'S OWN STORY. SEE THE LEGENDARY CONFRONTATION BETWEEN THE DREADED ELSWORTH WORM AND THE FRAIL MAIDEN. SEE TERRIFIED VILLAGERS AND MARAUDING DANES. SEE CERIDWEN MARTYRED FOR HER FAITH. OUR TOWN HAS SEEN NOTHING LIKE THIS IN A THOUSAND YEARS.

'Bit over the top, isn't it?' said Fliss. 'People'll be expecting a Hollywood epic and all they'll get is us, trolling about like wallies in a bunch of home-made costumes.'

Vicky chuckled. 'Doesn't matter, Fliss. They'll love it anyway. They always do when kids're performing. It's like the infants' nativity play where someone forgets her lines or bursts out crying or goes wandering offstage looking for Mummy. The teacher's going ape-shape thinking the whole thing's ruined, but it isn't, because the mums and dads think it's really cute. They've seen the play fifty times before anyway, and it's the things that go wrong that make it interesting.'

'Hmm.' Fliss wasn't entirely convinced. 'We're not infants, Vicky. You heard what Mr Hepworth said. The whole town'll be watching us. It's the last thing, you see – the climax of the Festival. It's a big responsibility and it scares me.'

They moved on, strolling in a great circle round the town centre till they found themselves outside Butterfield's once more.

'Another Coke?' suggested Fliss.

Vicky shook her head. 'I'd better go. We're off somewhere in the car this aft – some garden centre or something, and I'll have to get changed. What you gonna do – find that lad you fancied?'

'Which lad?' Fliss looked indignant. 'I don't fancy

anyone. I thought I'd walk round the supermarket – get a choc bar or something.'

Vicky grinned. 'I'll believe you. Thousands wouldn't. You around tomorrow?'

Fliss shrugged. 'Dunno. Depends what the wrinklies're up to. I'll give you a ring.'

Vicky departed and Fliss went into Butterfield's. It was hot and busy and she knew she'd spend half her time being jostled and the other half dodging trolleys, but then nothing's much fun by yourself and it was too early to go home. If she'd known what was about to happen among those crowded aisles, she'd have gone home anyway.

CHAPTER TWENTY-NINE

While Fliss and Vicky were reading the poster outside St Ceridwen's, Gary and the others were arguing in Trot's garage, which had become a sort of headquarters for them. This was where they stowed the pieces of the worm, and where they usually met. It was a big garage with plenty of space to spare even when the Trotters' Astra was in it, as it was now.

'I still say let's frighten some people,' insisted Gary. 'We all know how great we felt after we did it to old Ackroyd.'

'Yes,' said Lisa, 'but that was at night, and in a quiet spot. Going downtown in broad daylight's another matter. We'd get arrested.'

'It was you got in trouble for being out late,'

countered Trot. 'So Saturday morning should be just the job, right?'

'Yes,' put in Ellie-May, 'but what about the police, Trot? Wouldn't we be disturbing the peace or something?'

'Would we heck! Listen – Gary and me aren't stupid. We've got it all worked out. You know the other week, when the bookshop did that promo on kids' books?'

Ellie-May looked at him. 'Yes – what about it?'

'Well – they had guys dressed up, didn't they? There was a bogeyman, a puppy and an owl, all walking up and down the street in front of the shop. Did they get arrested?'

'Well no, but they were advertising something, weren't they?'

'Exactly!' Trot smiled. 'And so are we. If anybody asks, we're advertising our play, right?'

Ellie-May shook her head. 'I'm not sure, Trot. I don't know if we'd get away with it.'

''Course we would. And anyway, nobody's going to ask. Come on.'

Fliss was making her way towards the checkout with a three-pack of Snickers in her basket. The narrow aisle was thronged with trolley-pushing shoppers and their children. Just in front of Fliss, a kneeling youth was taking tins of peas from a trolley and stacking them on a shelf. The trolley blocked off half of the gangway, creating a bottleneck into which impatient customers were funnelled, pushing and shoving one another in their eagerness to progress.

Fliss was being swept towards this bottleneck and wishing she'd gone home with Vicky, when she became aware of some sort of commotion between the checkout line and one of the exits. She couldn't see very well because the stacked goods on the shelves were higher than she was and because people were craning to see, but there seemed to be violent movement in the crowd over there and she could hear exclamations of anger or maybe surprise.

Seconds later, the forward momentum of the crowd she was in ceased. For a moment, Fliss and those about her stood absolutely still. Then somebody screamed and the surge went violently into reverse as those at the front recoiled from whatever it was they could see. Fliss back-pedalled desperately as a tidal wave of shoppers threatened to overwhelm her. To her left, an old lady cried out and toppled, clawing at a pyramid of cans in a useless bid to stay on her feet. The pyramid collapsed, pelting the woman with cans as she fell. Other shoppers, skidding and stumbling through scattered cans, abandoned their trolleys, which became rolling barriers against those who came after. A child fell and was snatched by its mother from the jaws of certain death.

Fliss turned and fought her way to the top of the aisle where she clung to a freezer-cabinet. Bodies cannoned into her, threatening to sweep her away, but she hung on, and as she clung there, limpet-like, she saw the worm. It was coming along the walkway between the tops of the aisles and the fixtures which lined the back wall of the store. It was moving quite rapidly for a thing its size, scattering shoppers as it came. It passed within a metre of her, heading for the last, wide aisle which would lead it back to the end of the checkout line. Fliss watched as it swung round the bend, dragging its iridescent green tail, and disappeared from view. Then she turned with a moan and threw up over a hill of turkey parts.

CHAPTER THIRTY

Stan Morris had the biggest milk round in Elsworth. Seven days a week he was up at four-thirty and out delivering by five, and he'd work till ten at night, loading up his float for the next day. He followed this punishing routine the year round except for two weeks each January when he took Mrs Morris off to Florida. Renowned throughout the town for his addiction to hard work (some called him a workaholic), Stan would never win any prizes for the size of his imagination. In all of his forty-six years he had never seen a ghost or a UFO or a fairy and he never expected to, and he felt only scorn for those who claimed they had. So when a dragon crossed the road in front of the float at five-fifteen that Sunday morning, it came as a bit of a shock. He braked hard, causing the stack of crates on the flatbed to hit the back of his cab, and sat staring at the gap in the fence through which the apparition had vanished.

Mebbe the wife's right, he told himself. P'raps I have been working too hard. When a man starts seeing things it's time to slow down a bit.

Stan recovered his composure after a few minutes and drove on, and by eight o'clock he'd convinced himself he'd seen nothing unusual. Not that morning, nor any other morning of his life. For Stan, the unusual was deeply suspect and probably didn't exist.

Trot closed the garage door as silently as possible and tiptoed into the house. It was still only six o'clock. As far as he could tell, nobody had stirred. He crept upstairs and into the room he shared with his eight-year-old brother. As he eased the door closed, Jonathan rolled over in his bed and mumbled, 'Hnnn – where you been, David?'

'Sssh!' Trot pressed a finger to his lips. 'Mum and Dad are still sleeping, kiddo. It's early. I went to the bathroom, that's all.'

'Hmmm. OK.' The child rolled over again, pulled the duvet up around his ears and went back to sleep. Trot sat down on his own bed and bent forward to unfasten the laces of his Nikes, grinning as he did so.

Brilliant. It was brilliant. I'd give a million quid to see that wassock's face when he looks out the window.

The wassock Trot referred to was Percy Waterhouse, the Park Keeper, who was forever chasing teenagers, including Gary and himself, away from the kids' playground. Most teenagers still have a bit of the kid in them and they like the occasional swing or go on the roundabout and there was no harm in it that Trot could see, but Percy didn't agree. Big lugs, he called them, shouting and shaking his stick. 'Gerrawayfromthereyabiglugs!' What was a lug anyway, and who'd call their kid Percy, for crying out loud?

Anyway. Trot kicked off his trainers and stretched out on the bed with his hands behind his head, smiling at the ceiling. We paid him out this time, that's for sure. I wish I could be there when he sees what's left of his tulips. He'll go ape-shape. Cry in his cornflakes. He'll call the police but they'll not catch us.

A little voice in Trot's head told him that what they'd done was wrong, but that only served to broaden his grin. Wrong? Of course it was wrong. That was the whole point. He and his friends were discovering that doing wrong was fun. Oh, there was fear – a nagging, niggling fear behind the euphoria, which had little to do with the police and everything to do with the fact that, inside the worm, the four of them became one, in ways which Trot preferred not to think about. They saw through Gary's eyes, didn't they? Danced to his tune, submerged their minds in his, but so what? The kick was awesome, and afterwards they were their old selves again, so that was all right, wasn't it?

Well, wasn't it?

CHAPTER THIRTY-ONE

Ellie-May Sunderland's sister was away at college, so there was nobody to wake and ask her where she'd been when she slipped into her room. The Sunderlands always slept late on Sundays, so once she had her door closed she was safe. She should have been able to sleep, but for some reason she couldn't. She got undressed and slid in between the sheets, but then she just lay there thinking. She thought about how excited it made her feel to get into the worm with the others – how wonderful to be part of that invincible team. She remembered yesterday in Butterfield's – how people scattered at their approach. Their cries. The expressions of fear and disbelief on their faces. A part of her – some part she paid no

attention to because she didn't want to – kept asking
how you could see the expressions on people's faces
when you're the end bit of a worm. Deep, deep
down, she knew that something was happening to
the four of them. Something awful. Trouble was,
the excitement was so intense she didn't want it to
stop, and so she tried not to think about it. Instead,
she thought about the Park Keeper's tulips.

Terrific tulips they were. White and yellow, purple
and scarlet, all round the Park Keeper's house – a
rippling sea of colour with the house in the middle
like a galleon. Every spring they were there, and
people would make a detour in their journeys across
the park to look at them. Ellie-May could remember
when she was very young, being taken by her mother
to see the flowers. How tall they'd seemed on their
long stems – half as tall as Ellie-May herself. Nobody
grew tulips like Percy Waterhouse, and he was proud
of them.

Not now though. Not this year. It's amazing
what eight busy trainers can do to a bed of tulips
in the space of a couple of minutes. When Ellie-May
closed her eyes she could see what they'd done. She
could see the blooms lying bruised and broken on the
trampled earth, their lovely petals crushed and stained
with soil. She could see the torn leaves, the snapped-
off stems tilted drunkenly one against another like the
masts of a wrecked armada. She could see all of this

when she closed her eyes, as though the backs of her eyelids were a screen on which a video played, and it didn't make her feel good. She'd felt good while they were doing it. Then, her excitement had been intense, exhilarating. She'd laughed and whooped as she stomped and trampled, laying waste in seconds what had taken months to create. It had imparted a sense of power, a feeling that ancient wrongs were being avenged.

But now she only felt sad. Sad and frightened. What she and the others had done was wrong. She knew that now. Wrong, and stupid. Turning beauty into ugliness. Joy into tears. Good into evil. She thought of resigning her part – of giving up her place in the worm – but even as she thought about it, she knew she wouldn't. It was too wonderful, that buzz – that overwhelming wave of excitement, that sense of power. For some reason Ronnie Millhouse came tottering into her mind. Ronnie the drunk, who couldn't give up the thing which was destroying him. I'm hooked, she thought, just like Ronnie. The notion appalled her, but there it was.

'We do the most awful things,' she murmured aloud. 'Roll on the next time.'

CHAPTER THIRTY-TWO

'Well – what d'you reckon?' Percy Waterhouse looked at the Detective Constable. It was ten o'clock Sunday morning and the two men were standing among the Park Keeper's vandalized flowerbeds. 'It was obviously kids – it always is, but which kids? Have they left any clues?'

The policeman shook his head. 'That's what I'd have said, Sir. Kids. In fact, I'd have bet on it, but it seems I'd have been wrong on this occasion.'

'What – you mean adults did this? But why, in heaven's name? It's so senseless.'

The detective shook his head again. 'It wasn't adults either, Sir, as far as I can tell. It appears to be the work of some sort of animal.'

'Animal?' cried Percy. 'That's absolutely impossible. What animal would work its way systematically
round a garden, breaking every single bloom? I don't
believe it.'

'Well, Sir, I wouldn't have believed it myself,
but there are no human footprints that I can find.
Not one.'

'But you found animal prints?'

'Oh yes, Sir. Everywhere.'

'And what was it – a dog? A pack of dogs? What?'

'I don't know, Sir. Not yet. I'd like a veterinarian
to look at them before making any comment.'

'Will you show me some of these prints? I think
I can recognize dog prints without having to ask a
vet.'

'Certainly, Sir. Look here.' The Detective Constable stooped and pushed some bruised stems aside
with his palm.

Percy Waterhouse squatted and peered at the trampled soil. What he saw made him draw breath sharply.
'Good lord!' he gasped. 'What on earth made that?'

The policeman withdrew his hand and straightened
up. 'What indeed, Sir. D'you see now why I'd like an
expert opinion?'

'Yes I do. It's amazing.'

'Did your wife or yourself hear anything during
the night, Sir?'

'No. Not a thing. I knew nothing about this till

I opened the bedroom curtains and saw the mess. I assumed it was a straight case of teenage vandalism and rang the police.'

'OK, Sir – I think that's all for now. I'm going to leave a uniformed officer here to see that the ground remains undisturbed till the veterinarian's had a look at it. Will you be at home most of the day, Sir?'

'Oh, yes. At home, or patrolling the park.'

'Then I'll be in touch. Goodbye, Sir.'

'Goodbye, Constable.'

Percy stood gazing at the ruins of his garden. As he did so, he became aware that he was not alone. He turned and found himself looking at the pathetic figure of Ronnie Millhouse. The drunk was standing at the edge of the public footpath, regarding the Park Keeper through red-rimmed, watery eyes. As Percy turned and saw him, he nodded his unkempt head at the smashed flowers. 'Shame.'

'Yes.' Percy felt a stab of irritation. What did the town drunk know about tulips? What could he possibly care? He was probably about to cadge fifty pence or something.

'I reckon that there dragon done it.'

Dragon? Percy frowned. What was the idiot on about? He glared at Ronnie. 'What are you talking about?'

Ronnie gazed earnestly at the Park Keeper, who had spoken sharply, but whom Ronnie knew to be a good man. He knew that Percy knew he used the old bandstand for sleeping in, and that he could have kicked him out if he felt like it, but he didn't. Not only did Percy let him stay, but he sometimes left a bit of grub in a plastic bag for him to find. He never came when Ronnie was there, and if asked he'd have denied feeding the drunk, but Ronnie knew. When you're as alone as Ronnie, you develop a sharp nose for a friend.

'The dragon,' he repeated. 'I seen 'im t'other night, up the top path. Long he were, and green.'

Percy smiled faintly in spite of his grief. 'Not pink, then, Ronnie?'

Ronnie shook his head. 'Green. I hid behind a tree till he'd gone. I reckon it was 'im done this, Mister.'

'Well.' The Keeper smiled again. 'It's as good a theory as any I've heard up to now, Ronnie.' He smelled bacon frying and turned towards the house.

The drunk called after him. 'I'm right, Mister, you see if I'm not.'

Percy waved a hand without turning. 'Thanks, Ronnie. I'll bear it in mind.' He went in to his eggs and bacon, wondering briefly what Ronnie's breakfast would be, and when. As for the poor chap's dragon story, Percy had forgotten it before he closed the door.

CHAPTER THIRTY-THREE

Percy Waterhouse wasn't the only one calling the police that Sunday morning. Len Butterfield had spent a sleepless night wondering who it was who'd caused chaos in his supermarket the day before. He hadn't been there when it happened, but his manager had called him and he'd arrived before the staff had done much clearing up.

The scene which greeted him had made him very angry. The place looked as though a bomb had gone off inside it. Shelves were down. Cans and packets littered the aisles. Smashed bottles lay everywhere, their sticky, multicoloured contents spilled across the tiles. Abandoned trolleys stood with their tyres in this congealing goo. And worst of all, he'd found himself surrounded by a knot of irate customers who had been awaiting his arrival. Some of these customers had cuts and bruises to show him. Others displayed articles of their clothing torn, or decorated with globs of bleach, jam, mustard pickle and yeast extract. All of these people preferred their clothes the way they'd been before, and threatened to sue Len Butterfield for the cost of cleaning, repairing or replacing them. They seemed to think he was to blame for what had happened – they thought the monster or whatever it was had been some sort of publicity stunt gone wrong.

He tried to tell them it wasn't – that he knew nothing about it – but they were in no mood to listen. They'd all seen the weird collection of creatures outside the bookshop the other day. They knew traders would pull practically any sort of silly stunt to draw attention to their businesses, and it seemed obvious to them that Len Butterfield's stunt had simply got out of hand.

He'd smoothed it over in the end – made promises, given undertakings, and the customers had departed more or less satisfied. Some of them wouldn't be back though, and what with one thing and another, this stupid prank by persons unknown was set to cause Len a lot of unwanted hassle.

His manager had called the police and they'd arrived while Len was striving to soothe his customers.

They'd taken statements and poked about a bit, but they hadn't seemed all that interested and Len had suspected his little spot of bother wasn't serious enough for them. Nobody was dead or in hospital, nothing had been stolen, and any loss would probably be covered by insurance.

So at eight-thirty Sunday morning, after a restless night and with a pounding head, Len rang the station to demand a progress report. The sergeant at the other end was polite, but not helpful. Investigations had been made, and were continuing. There had been no significant developments so far, and if there were developments, Len would be informed. At this stage, said the sergeant, they were treating the matter as a prank, probably by children, which had got out of hand.

Which means, muttered Len, when he'd hung up and was making coffee, that you don't intend doing anything about it. You'll stick the statements in a file, shove the file in a drawer and forget about it. But I won't. I won't forget. I'll do my own investigating, and woe betide whoever did this when I get hold of them.

CHAPTER THIRTY-FOUR

It wasn't until seven in the evening that the Detective Constable got back to Percy Waterhouse. During the afternoon the Keeper had seen a woman he'd assumed was the veterinarian, squatting among the remains of his tulips with some sort of measuring device, but he hadn't gone out. He was too depressed to feel like talking, and anyway she was working for the police and probably wouldn't have told him anything. He'd watched through the window till the woman finished whatever it was she was doing and left with the uniformed constable, and he'd almost given up hope of hearing anything that day when he answered a knock on the door and found the detective on the step.

'Evening, Sir. May I come in?'

'Of course.' Percy stepped back to admit him, then closed the door and showed him into the sitting-room. 'Have a seat. Coffee?'

'Oh, no thanks. I had one at the station.' He smiled apologetically. 'There isn't a lot I can tell you, actually.'

'What did the vet say?'

The policeman shrugged. 'Not a lot. Reckoned the prints were like nothing she'd seen. Said they might have been made by a very large reptile but were far more likely to be an elaborate hoax.'

'Hoax?' cried Percy. 'Why would anyone want to pull a hoax like that – and how would they do it?'

The detective shook his head. 'I've no idea, Sir. We checked to see whether any large reptiles have been reported missing. They haven't, though we're continuing to monitor that. And we've searched the park. Oh, and I think I should tell you that while we were doing that, Jimmy Lee came sniffing around.'

'Jimmy Lee?'

'Yes, you know – chap from the local rag. Reporter. Nose like a ferret and features to match. Anyway, we had to tell him something so we gave him the elaborate-hoax line.' He grimaced. 'No doubt there'll be a piece in the *Star* about it. Thought I'd warn you.'

'Yes, thanks. And you really believe it was a hoax?'

'It's the likeliest explanation, Sir. We don't get a lot of large reptiles in Elsworth and anyway, what sort of reptile would do this sort of damage to a garden? Though as I said, we're still checking for possible escapes.'

'Well.' The Keeper gazed glumly out of the window. 'All I can say is, if it was a hoax I hope you catch the hoaxer. Oh, and by the way, there's at least one chap who'd go along with the large reptile theory.'

'Who, Sir?'

Percy smiled. 'Ronnie Millhouse. Swears he saw a dragon in the park the other night.'

'Yes, well.' The detective smiled too. 'Ronnie's got a whole menagerie of creatures inside his head, and they're all pink.'

'Not this one. Green, he reckons. Sure you won't have that coffee?'

'No thanks.' The policeman stood up. 'I'd better be off, Sir. We'll keep you informed if there are developments.'

'Thanks.' They walked to the door and Percy let his visitor out.

'G'night, Sir.'

''Night, Constable.'

He stood for a while, staring morosely at what was left of his garden. Then he sighed deeply, turned, and went inside.

CHAPTER THIRTY-FIVE

Jimmy Lee's scoop had broken too late for Monday's *Star*, so the people of Elsworth knew nothing of the dragon-and-tulip affair as the children of Bottomtop Middle streamed into school that Monday morning. Year Eight had planned no rehearsal for today – Mrs Evans had told them there was such a thing as being over-rehearsed – but in the event they had to sacrifice double English and get into their costumes because the vicar arrived during registration to ask how things were progressing.

They did it on the field, it went smoothly and the Reverend Toby East was impressed. When it was over – when Gemma Carlisle, the Viking Chief, had dragged Ceridwen off to her martyrdom between the goal-posts – he applauded. He actually stood there on the touchline with a smile on his face and clapped. Mrs Evans, who had stood beside him throughout the performance, clapped too. She felt she ought to, since Mr East had given the lead. He turned to her, beaming. 'Splendid!' he cried. 'Isn't it absolutely splendid, Mrs Evans?'

'Oh yes,' smiled the teacher, who would rather have been taking the double English lesson she'd prepared. 'Our Year Eight is a very able group, Mr East.' The children, who had heard the vicar's enthusiastic remark, came trooping across the field wearing bits of costume and smug grins. Even Fliss was smiling. Nothing unpleasant had happened and she was feeling better.

The vicar beamed at her. 'A fine Ceridwen, my dear – serene and lovely as the saint herself if I may say so.'

Fliss dropped her eyes, felt herself blush and murmured, 'Thank you.' She wished he hadn't singled her out for praise – Gary Bazzard wouldn't like it, and she was anxious that he should be propitiated till after Saturday.

The vicar said something to Mrs Evans, who nodded. He turned to the children. 'Would the four children who play the worm please remain here for a moment?'

Fliss saw that his smile had gone and felt a spasm of unease. Please don't stir them up, she thought. It's me that's got to face them on Saturday.

Mrs Evans touched her shoulder. 'Come along, Felicity.' The class was making its way back into school. Fliss followed, hoping the vicar had kept the four to praise them; knowing he had not.

'Now.' The vicar regarded the quartet sternly. 'I'm going to ask the four of you a question, and I want you to answer me truthfully. Is that understood?'

Gary Bazzard nodded. The others mumbled, 'Sir.'

'Where were you at ten past eleven last Saturday morning?'

'At my granny's,' said Gary at once. 'You can ask my mam.' The others looked at him and said nothing.

The vicar sighed. 'And you, young lady – where were you?'

Lisa looked from Gary to the vicar and back again, biting her lip. 'It's no use, Gary,' she said. 'He knows.'

'Yes.' The vicar's tone was icy. 'He knows, but he's waiting to hear it from you. Where were you?'

'Butterfield's,' mumbled Lisa.

'Supermarket,' said Ellie-May.

'In costume,' admitted Trot.

'Thank you,' said the vicar quietly. 'You might be interested to know that one of the shoppers in Butterfield's that morning was my wife. Your antic upset her quite badly, but unlike everybody else she knew about the play and realized where the monster must have come from.' He frowned. 'I suppose you know what a wicked thing it was that you did?' Nobody answered. 'You know, don't you, that your silly prank might have had disastrous consequences? Somebody frail – a weak heart perhaps, and they might have died. Did you think about that? Did you consider the possibility of somebody being trampled, crushed – somebody's baby? Did you think at all before you did what you did?'

Lisa sniffled. 'No, Sir.'

'No, Sir.'

The vicar gazed at them. 'Why did you do it, eh? Whatever possessed you to do such a thing?'

'Possessed?' Gary glanced sharply at the vicar. 'Nothing possessed us, Sir. It was a stunt. A publicity stunt, to advertise our play. We thought it was a good idea, Sir, that's all.'

The vicar looked at the boy. 'Your idea, was it?'

'Yes, Sir.'

'Well, it was not a good one, Gary. Far from it. People were injured. Frightened. Property was damaged. And there was nothing to connect the event in people's minds with your play. If, as you say, it was a publicity stunt, it was poorly thought out and brutally executed, and I'm ashamed of you. Your classmates have worked hard to produce an outstanding presentation, and the four of you have let them down with this act of – of vandalism. Do you know that the police are involved?'

'Police?' Trot looked scared.

'Of course.' The vicar sighed again. 'Oh, it's all right. You needn't worry. I'll go to them. Tell them it was a publicity stunt gone wrong. I'll talk to Mr Butterfield too. It will be all right. But I want you to promise me that you'll never ever do anything of the sort again. Do I have your promise?'

They nodded. 'Yes, Sir.'

'Good. Then we'll say no more about it. That's a very fine costume you've constructed. Most realistic. Keep on rehearsing, and good luck for Saturday.'

'Thank you, Sir.'

The vicar strode away, and the quartet walked slowly across the yard. 'What now?' asked Lisa. 'We've given our promise.'

'Our promise?' Gary kicked a stone. 'What're you – an infant? Our flipping promise!'

'He's the vicar, though. A promise to a vicar's sort of special, isn't it?'

'Oh, yes.' The boy grinned wolfishly. 'It's special all right. 'I'll get special pleasure out of breaking it, that's what's special about it.'

CHAPTER THIRTY-SIX

Tuesday morning, seven-thirty. The Morgans at breakfast. Mrs Morgan sips coffee. Mr Morgan hides behind the *Star*. All you can see of him is his fingers and the top of his balding head. Fliss takes the last slice of toast from the rack and begins to butter it. Her knife makes a scratchy sound on the toast. The *Star* is lowered slightly. Her father glares at her over the top of it. He doesn't say anything. He doesn't need to. 'Sorry,' murmurs Fliss. She butters more quietly. The *Star* rises to its former position. Silence, which Mr Morgan breaks with a scornful laugh. His wife and daughter glance up, waiting to know what's funny. Without lowering the paper, Mr Morgan begins to read aloud. Fliss

137

wonders how he knows they're listening.

'Park Keeper Percy Waterhouse called the police on Sunday morning when he found his formerly beautiful garden had been wrecked in the night. When the constabulary arrived at the scene, huge reptilian footprints were found all over the Keeper's tulip beds. A veterinarian who examined the prints dismissed them as a hoax, and a police spokesman told our reporter, "We don't get a lot of large reptiles in Elsworth." However, when our reporter spoke with Mr Ronnie Millhouse, a resident of the park, Mr Millhouse claimed to have seen a large dragon there only a few nights ago. Most people would doubtless be inclined to discount this evidence, but before doing so they ought perhaps to consider the following: Elsworth once played host to a very large reptile indeed. This reptile was no hoax – it ate people. The beast was never killed – it was simply banished to the fen. This was exactly one thousand years ago. This week the people of Elsworth are celebrating its banishment.

Prematurely—?'

Mr Morgan stops reading. The silence lasts several seconds.

'Go on,' says Mrs Morgan.

'That's it,' her husband tells her. 'There's no more.'

'What an odd story,' says Mrs Morgan.

'Damned silly if you ask me,' growls Mr Morgan. They both chuckle.

Fliss does not.

Because of the impromptu run-through for the vicar on Monday, Tuesday's rehearsal was cancelled. Fliss was glad. She couldn't get the *Star* story out of her head. Common sense told her that Gary and the others must have been on the rampage again, but would they dare do such a dreadful thing? And what about the prints? How had they managed those? She longed to ask Lisa but knew she mustn't. Lisa wouldn't tell her anyway. From time to time during that seemingly endless day she watched Lisa and the other three, hoping they'd give themselves away by some word or expression but, though the dragon story was the chief topic of playground conversation, she detected nothing which might indicate their guilt in the affair. She discussed it at lunchtime with Vicky, who said it couldn't have been them – the footprints would have been far too difficult to fake.

Nevertheless, Fliss worried. She worried all day at school, and all evening, moping around at home. Finally, at nine o'clock, she could stand it no longer. She should have been thinking about going to bed, but she knew she wouldn't sleep till she knew what had happened Saturday night in the park. Her parents exchanged glances when their daughter announced that she fancied a pizza takeaway and got into her jacket, but the take-away was only round the corner. 'Don't be long, dear,' was all her mother said, and her father chipped in with, 'And don't talk to any strange men.'

She reached Trot's gateway and hesitated. Suppose Trot and the others weren't here? She knew they met most evenings, but maybe not tonight. Well, she told herself, if they're not here I'll knock on the door and ask to see Trot. I'll tackle him head on – ask him straight out whether he and the others wrecked the Keeper's flowers, and how they made the prints. It might even be easier if he's by himself. I'll swear not to tell on them, if only he'll set my mind at rest. Yes. That's what I'll do. I sort of hope he is alone. She took a deep breath and strode up the path.

They weren't there. The garage door was up, but the place was empty. There was no car. And when she crept inside and looked around, there was no worm either. The costume had gone.

She tried the house. There were lights on inside, but nobody answered her knock. With the car gone, it was likely the Trotters were out for the evening. And with the worm gone, it was likely the foursome were out for the evening too.

Where? As she turned and hurried down the path, Fliss felt a tingling in the nape of her neck. She half ran along the road, glancing back from time to time to make sure nothing was following her. She was so scared she almost forgot to get her pizza, and when she got it home she couldn't eat it. She shot it into the pedal-bin and ran upstairs to her room. Her parents exchanged glances again.

'Hormones,' said Mrs Morgan.

'Aaah,' said Mr Morgan.

CHAPTER THIRTY-SEVEN

If there was one thing Jimmy Lee enjoyed more than sniffing out a good story, it was his pigeons. They were racing pigeons and Jimmy had twenty of them, not counting squabs. He kept them in a loft he'd built himself, on an allotment opposite his house. This allotment was on the same block as Hughie Ackroyd's, and the two men were on nodding terms. Hughie didn't care for pigeons and Jimmy wasn't interested in growing vegetables, but they did have one thing in common – they were both worried about the kids who hung around the abandoned greenhouse. Bored kids often got up to mischief, and a neat garden or a well-ordered pigeon loft might well act as magnets to acts of casual vandalism.

So when Jimmy looked out of his window that Wednesday morning and saw that the door of the loft was swinging in the breeze, his first thought was that his birds had had a visit from those flipping kids. Fearful for the welfare of the squabs, he pulled on some clothes and hurried across the road, to find that the situation was very much worse than he had feared.

The first thing he noticed was the smell. It was a pungent smell and Jimmy recognized it. It was the smell of burnt feathers, and he could smell it before he reached the loft. He hurried forward, cursing under his breath, and cried out in horror and disbelief at the sight which met his eyes.

They'd had the place on fire. The structure itself hadn't burned, but the inside walls were scorched as though somebody had stood in the doorway and discharged a flame-thrower. Dead birds littered the floor, their plumage blasted off. Charred feed-bags spilled their contents among the corpses, and inside the nest-boxes his precious squabs lay roasted on beds of blackened straw. So intense had been the heat from whatever it was the vandals had used, that the loft's window had cracked across two of its four panes. A quick count told Jimmy that not all of his birds had died, but of the survivors there was no trace.

He was trudging back, intent on calling the police, when he saw the footprint. There was only one, in

143

a patch of soft earth near the gate. Jimmy squatted, tracing its outline with a finger. It was big – maybe thirty centimetres across, and it had been made by something heavy because the depression was at least four centimetres deep. In fact, it was exactly like the prints the police had shown him in the park yesterday.

'Some hoax,' he muttered, straightening up, wiping soil from his finger on his jeans. 'Some rotten hoax.' Tears of grief and rage pricked his eyes. He kicked a stone viciously with the toe of his trainer and strode towards the house.

CHAPTER THIRTY-EIGHT

'Mr Bazzard?'

'Yes.'

'We're police officers. You have a son, I believe
– Gary, isn't it?'

'That's right. Why – what's he done?'

'I haven't said he's done anything, Sir. We'd like
a word though. Is he in?'

'Yes. Upstairs. You'd better come in.'

It was six o'clock Wednesday evening. Gary's
mother was out. She worked the evening shift at the
biscuit factory. Her husband was glad. She'd have a
fit if she was here, he told himself. Police asking after
our Gary. He went to the foot of the stairs.

'Gary!'

'What?'

'Someone to see you.'

'Who?'

'Police.' The loud music which had been issuing from his son's room ceased abruptly and Gary peered over the bannister.

'Police? For me? Why?'

One of the two officers looked up at him. 'Come on down, son, and we'll tell you.'

Gary descended like a man on his way to be hanged. His father led the officers into the front room. 'Sit down if you like. Do I stay or what?'

'Stay if you wish, Sir. This won't take long.' Both officers remained standing. They looked at Gary. 'You're at Bottomtop Middle, aren't you, son?'

Gary nodded warily. ''Sright.'

'And you're involved in a play. Part of the Festival.'

'Yes.'

'You're part of a dragon thing, aren't you?'

'The Elsworth Worm. I'm the head.'

'A remarkably realistic beast, by all accounts. Where is it?'

'Sorry?'

'Where is it – the prop, costume, whatever you call it?'

'Oh it's at Trot's place. David Trotter's. He's part of it too.'

146

'You keep it at your friend's house?'

'Yes.'

'And what do you do – do you get into it sometimes and practise for the play?'

'We rehearse, yes.'

'Who's we, son? How many of you are there?'

'Four.'

'Names?'

'David Trotter. Ellie-May Sunderland. Lisa Watmough. And me.'

'And you rehearse together?'

'We have to. It's not easy, moving together and all that, when only one can see.'

'I can imagine. Where do you rehearse?'

'Trot's garden. The street. Anywhere, really.'

'Hughie Ackroyd's allotment, perhaps? Butterfield's supermarket?'

'I – we don't use anybody's allotment. We did go round the supermarket last Saturday, but that was for publicity.'

'Publicity?'

'For the play. It was sort of an advert.'

'Pretty violent advert, son.'

'I know. The vicar told us off. It like – got out of hand.'

'I'll say it did. And what about the stunt in the park, and the one with Jimmy Lee's pigeons? Did they get out of hand too?'

'I don't know what you mean. What stunt in the park? What pigeons? I don't know anything about any pigeons.'

'Tell me how you do the footprints, son.'

'Footprints?' Gary looked bewildered. 'I don't understand.' He appealed to his father. 'Dad – I don't know what he's on about.'

Mr Bazzard looked at the officer. 'What's this all about, Officer? Why are you questioning my son?'

The officer told him. When he'd finished, Mr Bazzard frowned. 'And you think my son'd do something like that? Destroy somebody's garden? Burn up a man's pigeons? He's just a kid, for Pete's sake.'

'We're not accusing him, Mr Bazzard. We're simply making enquiries.' He turned to Gary. 'David Trotter – you know his address?'

'Sure. Thirty-three Baslow Grove. He'll only say the same as me.'

'I want to take a look at this worm of yours, son. See what sort of feet it's got.'

'Our feet. It's got our feet, that's all.'

'Then you've nothing to worry about, have you, son?'

They left. Gary told his father about the supermarket and reaffirmed the quartet's innocence in the other matters the officer had mentioned. Then he went up to his room, turned up the volume on his CD player, and lay on his bed wondering who had set the police on to him, and in what way the blabbermouth might be made to pay.

CHAPTER THIRTY-NINE

'Are you feeling all right, Fliss?' Thursday morning. Fliss playing with her cornflakes, watched by her mother who looks concerned. No, Mother, I'm not feeling all right actually. I'm tired, and I'm scared. Fliss doesn't say this, though it's the truth. It would lead to questions she'd rather not try to answer. She wants only one thing. She wants Saturday to come and go so normal life can resume. Till then she wants to be left alone. She forces a smile.

'I'm OK, Mum.' I'm not though. I fret, I dream, I fret some more. Things are happening. Frightening things. Things there are no words for. And the ship sails on.

This is the ship. This house, the street outside,

people's lives. The good ship Elsworth, sailing towards disaster while everybody dances. The only one who knows is me and I'm just the cabin boy. If I tried to tell them, they'd laugh.

'You don't look too well, dear. Didn't you sleep?'

'A bit. It's just the play, Mum. I'll be fine when we've done it.'

'Hey, listen to this.' Dad, from behind his paper. How do I know it's Dad, thinks Fliss. It could be anyone with thin hair and thick fingers. Someone different every day.

'What is it, dear?' her mother enquires.

Dad grunts. 'Another dragon story, but not so joky this time. Listen.' He reads out the piece Jimmy Lee has written. The one about his pigeons, and how a hoax can go too far. Fliss puts down her spoon and gazes into her bowl. Her fingers pluck at a corner of the tablecloth, pleating, smoothing, pleating. Nobody notices. Dad comes to the last bit – the bit where Jimmy says that somewhere in Elsworth there's a dangerous person – perhaps even a madman – and that the sooner he's caught the better. He stops reading and lowers the paper. His wife sighs, shakes her head and murmurs, 'Some people.' Fliss stops playing with the tablecloth and picks up her spoon. Her hand trembles and she feels sick. All she can think is, Tuesday night. It happened Tuesday night and they were out. I know, because I went round and there was nobody there. It was them, and on Saturday I've got to face them. Or it.

On Saturday, I'm the pigeon.

CHAPTER FORTY

Mrs Watmough shook her head and clucked into her coffee. Her husband had gone to catch his London train and she was lingering over a second cup, the *Star* spread before her on the table.

'What is it, Mum?' Lisa, clearing breakfast things, leaned over her mother's shoulder.

'This.' The woman tapped an item with a fingernail and Lisa read the headline. PRIZE BIRDS BLASTED IN LOFT RAID.

'Where was this?' she asked. 'What happened?'

Her mother read out the piece, then shook her head as before. 'There's some wicked folks about,' she sighed. 'They want locking up, or worse.'

Lisa picked up the toast rack and butter dish. 'It's awful,' she said. 'Poor little things.' She turned away to hide her shining eyes, remembering the reek of burning feathers.

CHAPTER FORTY-ONE

Fliss set off as though going to school, but she didn't go. She'd remembered something her dad had read out on Tuesday about Ronnie Millhouse. Ronnie had claimed he'd seen a dragon in the park. What did he mean? Did he mean a real dragon, or people dressed up? Was he drunk when he saw it? People said Ronnie Millhouse was permanently drunk. Well, OK, but did that mean he wouldn't be able to tell a real dragon from a pretend one? There was only one way to find out.

She was desperate. If she hadn't been, she wouldn't have dreamt of approaching Ronnie Millhouse. She was afraid of him. All the kids were, because of the shouting. Most of the time, Ronnie was quiet, but now and then he'd go off his head and start shouting. Spit would fly from his lips on these occasions. He'd wave his arms about and shake his fists and the things he shouted made no sense. When he was like that, even adults gave him a wide berth.

She knew where he'd be at this time of day. He'd be at the bus station, cadging change from people going to work. And if he'd already left there, he'd be by the stall in the market where he went to buy a mug of tea and get warm.

He wasn't in the bus station, so Fliss hurried across to the market and found him by the tea stall with his big raw hands wrapped round a steaming mug. The stall owner was wiping off his counter but there were no other customers. Probably there wouldn't be till poor Ronnie moved on.

Fliss approached gingerly, praying that the drunk wouldn't start shouting. She didn't want people to see her talking to him. She was ashamed of herself for feeling like that but she couldn't help it. She drew near, smiling.

'Hello.'

'Hello, love. No school today?'

Relief washed over her. He sounded just like anybody else. She shook her head. 'I'm not well.' Which is true enough, she thought.

'Poor lass. Cuppa tea, is it?' For a moment she thought he was offering to buy her one, but when he made no move towards the counter she realized he wasn't, and was glad.

'No. I – I saw your name in the paper.'

'Oh aye? What'd it say then?'

'It said you saw a dragon in the park.'

'Dragon?' Ronnie's eyes clouded over and his face creased up with the effort of remembering. 'Oh, aye. The dragon.' He smiled ruefully. 'Nobody believed me.'

'I do.'

'Do you?' He grinned at her. 'Good for you. It's hard when you tell the gospel truth and nobody believes you.'

'I know.' She smiled. 'It happens to kids all the time. What was it like?'

'What was what like?'

'Your dragon, Mr Millhouse.'

''Twern't my dragon, love. I only saw it, that's all.'

'Yes, but what was it like? Was it kids dressed up?'

'What?' He glared. 'I thought you said you believed me?'

'Oh, I do!' She said this quickly, afraid he'd start shouting. 'I'd like to know what he looked like, that's all.'

'Well, he weren't nobody dressed up, I can tell you that. He were long and green and his head was up in the air. Red eyes, he had, and fire in his mouth.'

'Fire? Are you sure?'

''Course I'm sure. He passed me as close as you are now, and I felt the ground shudder from his footsteps. Scared I was, I can tell you.'

'And had you ever seen him before?'

Ronnie shook his shaggy head. 'No. Never before, never since, nor never want to neither.'

'Oh.' She'd learned nothing that was of comfort, and didn't know how to break off the conversation. She looked at her watch. 'I must get on.'

'Doctor's, is it?'

'What?'

'Doctor's – you being unwell and all?'

'Oh – yes, that's right. Ten past nine at the surgery. 'Bye, Mr Millhouse.'

'Mind how you go, love – there's some funny people about.' He glared about him at the early shoppers. 'I SAY, THERE'S SOME FUNNY PEOPLE ABOUT!'

Fliss turned and fled.

CHAPTER FORTY-TWO

It's impossible, surely? A real dragon. The actual Elsworth Worm. Ronnie Millhouse is a drunk, right? And drunks see things that aren't there – pink elephants and stuff.

So why did I go and see him? I wish I hadn't. A bunch of kids dressed up – even violent kids who hate you – is one thing. I can handle that. But the actual worm—

Ah, come on! What are you, Fliss Morgan – two years old? An infant, scared of the Big Bad Wolf?

There wasn't much point going to school till lunchtime, so to avoid being seen truanting she'd come to the park. People walk through parks, of course – especially on warm spring mornings like this one

159

– so Fliss had found a nice quiet spot where she could sit and think. The old bandstand stood in a forest of laurel and rhododendron gone wild, in a part of the park which was seldom visited except by kids at weekends and young couples in the evening. It was Ronnie Millhouse's bedroom, of course, but Fliss didn't know that. She thought the pile of old newspapers and bin liners under the bench must have been dumped there by some wally too idle to find a litter bin.

She was scared now. Really scared. The more she thought about recent events, the more convinced she became that something sinister was happening. All right, there might not be an actual worm – probably never had been; it was just a legend – but something had definitely happened to Lisa and the others. They'd changed. Before, they were just ordinary kids. Sure, they got up to mischief now and then like anybody else, but they'd never have dreamt of invading Butterfield's or trashing somebody's garden. And as for that awful thing with the pigeons— She shook her head. It wasn't them. It couldn't have been. Fire had been used. They'd never do that. They'd changed, but not that much.

And yet— She was thinking about last Saturday in Butterfield's. It was a bit hazy now – she'd sort of blotted it out – but what exactly had she seen? What had all those shoppers seen to make them panic as they did? Some kids dressed up? A papier-mâché head and eight trainers pounding the floor? Try as she might, Fliss couldn't remember those eight trainers. She thought she remembered something else. She thought she could remember four sturdy legs and gigantic birds' feet slapping the tiles. Birds' feet with long, crescent talons— But no – she must be mistaken. It must have been an hallucination – a shared hallucination. They happen. Hundreds of soldiers once thought they saw an angel hovering over the battlefield at a place called Mons. And the Indian rope trick – that was supposed to depend on spectators sharing the same hallucination.

It wasn't a convincing explanation. It didn't make her feel any better. A draught kept stirring the papers under the bench, making them rustle. Making her jump. It was a quiet spot she'd chosen. Too quiet. After a while she decided she'd rather risk being seen by somebody who knew her than remain in the bandstand. She got up and walked out into the sunshine, which failed to cheer her.

CHAPTER FORTY-THREE

Fliss approached the school gate at ten to one, and the first person she saw was Lisa. She dropped her eyes and made to pass her former friend without speaking, but Lisa said, 'We didn't expect to see you again till Monday. We thought you'd got the message.'

Fliss stopped. 'What message?'

Lisa sighed. 'The message we've been sending you for weeks now. Stay away. Don't play Ceridwen. Fall sick. Let Samantha do it instead. That's what understudies are for.'

Fliss gazed at her. 'I don't get it. What difference is it going to make whether I play the part, or Samantha? The ending'll be the same.'

'No it won't. If you play the part, something

162

terrible will happen to you. If you don't, it will happen to Samantha. That's the difference.'

'What is this terrible thing you say will happen, Lisa?'

Lisa sighed again. 'Have you heard about Jimmy Lee's pigeons?'

Fliss swallowed hard. 'Yes, I've heard. What about it?'

'We did that.'

'What? I don't believe you. You wouldn't burn little baby birds in their nests.'

'We wouldn't, but we did. And we wrecked Percy Waterhouse's tulips, and we got away with it too.' Lisa's eyes gleamed. 'The police came to Trot's. Searched.' She laughed. 'They wouldn't tell us what they were looking for, but we knew. They were looking for some gadget they thought we had for making footprints. They didn't find anything, of course, and Gary said, "Even if we did have a way of making prints, there'd still be our own footprints, wouldn't there? How would we get rid of them, Officer?" Lisa laughed again. 'Officer, he called him, in this very sarcastic voice.'

Fliss looked at her. 'And what's the answer, Lisa? How do you do it?'

Lisa's grin faded. She shivered. 'You wouldn't believe me if I told you,' she said. 'I can hardly believe it myself.'

'Try me.'

'No. Listen, for the last time. Something's started here which nobody can stop. I tried to tell you right at the beginning, remember? I said it was as if something was taking over, making things happen. Well, I was right, and now this thing's in control and none of us could stop even if we wanted to. You'd have to be inside the worm to understand, but you're outside and you're in the way, and that's not a smart place to be. We've been friends, Fliss, and that's why I'm warning you. Stand aside, or suffer the consequences. I can't make it any plainer than that.'

Fliss gazed at her. 'I don't believe you, Lisa. I don't understand some of what's happened but I think it's you and Gary and Trot and Ellie-May, playing some sort of game. You've done some cruel, stupid things to try to frighten me, but I don't believe you burned the pigeons. Somebody else did that, and you're just using it to make yourselves seem ruthless. I'm going to be there on Saturday, and that's where your game will have to stop because there's nothing special about the four of you, Lisa. Nothing. You're a bunch of kids, that's all, and once the play's over they'll scrap the costumes and that'll be that.' She spun on her heel and strode off down the driveway.

Lisa gazed after her. 'You're wrong, Fliss,' she murmured. 'You've no idea what you're up against, but you'll find out. Trouble is, by then it'll be too late.'

CHAPTER FORTY-FOUR

Friday afternoon had been set aside for a last full-dress rehearsal on the school field. They hadn't rehearsed Thursday afternoon, but the whole of Year Eight had gone with Mr Hepworth across town to get a look at the Festival Field. Sarah-Jane had taken notes and made sketches, so that everybody would know where to stand and how to move during the actual performance.

Lunchtime. For Year Eight this meant a quick bite, then off to the changing-rooms. They'd done it all before and things had generally gone well, but everybody was feeling nervous just the same. This was it. The final run-through. Next time they took the field, it would be the real thing.

Fliss hung back a bit when it was time to change. She wasn't in the early part of the play anyway, and she didn't particularly want to run into Lisa and Ellie-May. All they had to do was put on green tights – they'd get into the worm on the banking behind the goal-posts – so they shouldn't be long in the changing-room. She loitered in the yard till she saw them leave, then went in.

She'd hung the bridesmaid dress on a peg that morning so that any creases might drop out. As she approached it, she saw that somebody had fastened a small sheet of paper to the bodice with a pin. With hands that shook she pulled out the pin and smoothed the paper. It had been torn from a jotter, and somebody had scrawled a verse on it in pencil:

NEVER WORRY
SLEEP ALL DAY
NEVER GO TO SCHOOL
NEVER TIDY UP YOUR BEDROOM –
BEING DEAD IS COOL

She read it through twice. Whoever had written it had used block capitals so there was no handwriting to identify him, but Fliss knew who the poet was. She balled up the paper and flung it into a corner. 'Never give up, do you, Gary Bazzard?' she murmured. 'But you might as well, because here comes the bridesmaid.'

The rehearsal went perfectly. Mrs Evans and Mr Hepworth watched from the touchline as the worm terrorized the villagers. This was Year Eight's favourite bit, and it went on for some time. It never got boring though – the worm was wonderful to watch, and each of its victims had a different way of screaming. They watched as the beast came strutting from its fen to claim another life and found Ceridwen standing in its path. They thrilled in spite of themselves as the creature lunged, roaring, at its frail adversary, but they knew nothing of Fliss's relief when it brushed her dress, grew docile and slunk away.

The rest was easy. Gemma led her Vikings in a series of convincingly bloody raids on the village. More screaming. Having subjugated the villagers, Gemma demanded that they worship Viking gods. Ceridwen refused and was butchered. There was a brutal-looking axe and plenty of tomato ketchup, but no screaming. Year Eight had decided that saints don't scream.

If they'd been anywhere near Fliss at two o'clock Saturday morning, they'd have learned how wrong they were.

CHAPTER FORTY-FIVE

She awoke to utter darkness and a rank odour she could not at once identify. She was cold, and her bed seemed to have grown hard while she slept so that her back, bottom and heels felt bruised. Groaning softly, she tried to roll on to her side, but her right knee encountered an obstruction which prevented it bending. Puzzled, she flexed it again and felt the kneecap press against something which did not yield.

Unease stirred in her. She lifted an arm, and the hand struck something solid no more than a few centimetres above her face. A whimper constricted her throat. She groped frantically with both hands in the blackness, and the nails and knuckles scraped something smooth and hard. She tried to fling her arms wide, but her hands thudded into solid matter, producing a hollow sound and causing pain. As this pain ebbed, she recognized the smell which filled the darkness. It was the reek of wet earth.

She could hear voices. Children's voices, chanting in unison:

NEVER WORRY
SLEEP ALL DAY
NEVER GO TO SCHOOL
NEVER TIDY UP YOUR BEDROOM –
BEING DEAD IS COOL

and it was then that she knew she was in her grave.

Screaming, she shot bolt upright and nothing stopped her. The mattress gave under her hands and bottom, and the reek of earth faded. There were footfalls and a flood of glorious light and then she was clinging to her mother, sobbing and shaking and babbling something about a grave. Her mother rocked her and stroked her hair, but it was some time before Fliss grew quiet.

CHAPTER FORTY-SIX

Sunlight lay in dapples on her duvet when Fliss woke up. She knew she'd been dreaming, but could not remember her dream. It felt late. She rolled over, grabbed the clock on the bedside cabinet and gasped. Eleven. It was eleven o'clock. Practically everyone in Elsworth would be making their way to the Festival Field by now, ready for the afternoon's festivities. People would have been working since early morning, erecting stalls and stands, tents and booths. Hanging flags and bunting. Putting up signs and notices.

'Mum!' She sprang out of bed and began pulling on her clothes.

Her mother came hurrying up the stairs. 'Fliss

– are you all right, dear?'

Fliss nodded. 'Sure, but look at the time. Why didn't you wake me? You know we're doing the play today.'

'Of course I know, Fliss. It's at two o'clock. Your dad and I are ready, but there are three hours yet and we thought you ought to sleep on awhile after the dreadful night you had.'

'Did I have a dreadful night? I'm fine now.'

Her mother nodded. 'You certainly did, young woman. Two o'clock this morning, screaming your head off. You'd had a nightmare. Something involving a grave, from what I could make out. Don't you remember?'

'No. Well – vaguely. I was in my grave, I think, and somebody was singing.'

'You frightened me half to death, I know that. There's nothing worse than being woken in the middle of the night by a scream.'

'Sorry, Mum. I think I know what brought it on.'

Her mother nodded grimly. 'So do I, dear. It's this play. It's been worrying you for weeks. It's been like having a little stranger in the house, the way you've mooned and fretted. Not like you at all.'

Fliss nodded. 'I know.' And I'm still worried, she thought. More than worried. I'm scared. Not of Gary Bazzard and the others, though. No. Something else. Something'll happen today. Something that isn't in the script. I know it. I can feel it deep down, but I can't talk to you about it, Mum. Or Dad. You'd think I was barmy. No, it's something I've got to face by myself. Aloud, she said, 'Is my dress ready?'

Her mother nodded. 'I ironed it. Nobody'll notice the stain. Dad's put it on a hanger in the car.'

'Good. I mustn't forget my sword.' A plastic sword, she thought. What use will that be when it comes – whatever it is?

She tried to eat breakfast, but could manage only orange juice.

'You can't fight a dragon on that,' joked Dad. Fliss forced a smile.

And so it was that at a quarter to twelve on that sunny April Saturday, Fliss set out with her parents to face whatever it was that awaited her on Elsworth's Festival Field.

CHAPTER FORTY-SEVEN

Trot had been up and about since six. He'd woken at five-thirty, full to bursting with energy and anticipation. Unable to suppress this he'd slipped out of bed, dressed silently and let himself out of the house.

He spent nearly an hour tinkering with the worm. He tapped extra staples into the frame at points where wire and wood threatened to part company. He used superglue to fix a couple of loose teeth. He gave the fabric a vigorous brushing where it had picked up splashes of mud, and touched up the paintwork here and there on the head. He whistled as he worked, because he felt that today was going to turn out special for himself and his three friends. Today they'd leave something behind and start something new and nothing would ever be the same again.

At eight o'clock, Gary phoned. Was he ready? Was everything set? He sounded high, and told Trot that he'd phone Lisa and Ellie-May to make sure they were ready.

Ready for what? As he put down the phone, Trot felt a surge of dull fear. What was happening to them all? What was it they'd got into? How would it end? There were no answers to these questions. The fear was a part of the excitement – the sick kick he felt – and all Trot knew for certain was that the sensation was mounting and that he couldn't stop now if he tried.

It's going to be terrible, he moaned. The worst thing that anybody ever did. I don't know how we can even think of it.

Roll on two o'clock.

CHAPTER FORTY-EIGHT

At twenty to two, Fliss slipped away from her parents with the bridesmaid dress folded over her arm and the sword concealed inside it. The field, outside the roped-off central arena, was thronged with people, and she had to dodge and weave her way through them as she headed for the marquee in which she and the others would change. There were still a few picnickers, but most people had packed away lunch and were watching two clowns wobbling on unicycles around the oval of cropped grass which formed the arena, juggling burning torches. The marquee stood at one end of the arena and when Fliss reached it, most of the kids were there already.

'Here's our Ceridwen,' grinned Mr Hepworth as she ducked inside.

Mrs Evans smiled tightly. 'Just in time, Felicity. Hurry up and change now.'

The marquee was crowded with Vikings and villagers. Fliss glanced around till she spotted Gary and the others, but they were occupied with their costume and didn't glance her way. She swallowed hard, told herself not to be silly, and began to change.

She'd put on the dress and was buckling her white sandals when the vicar arrived. He said something to Mr Hepworth, who clapped his hands to get everybody's attention. 'Listen,' he said. Andrew Roberts continued practising his narrative on Barry Tune. The Deputy Head glared at him. 'When you're quite ready, Andrew Roberts.'

'Oops – sorry, Sir.'

Mr Hepworth sighed. 'The Reverend East has a few words to say to you all, so pay attention.'

The vicar beamed. 'Good afternoon, everybody. In a minute or two I shall go to the podium to announce the commencement of your splendid production, but I thought I'd drop by here just to say how much I appreciate all the hard work you people have put in in the three weeks since Easter, and to tell you how much I'm looking forward to your performance.' He smiled. 'Good luck, everyone.'

'Thank you, Sir,' chorused Year Eight, high on

adrenaline. The vicar walked out into the sunshine.

Mrs Evans cleared her throat. 'Right, Year Eight, this is it – your big moment. You've worked terrifically hard and everything's fine, so don't worry. Go out there and enjoy yourselves, and the whole town will enjoy you too.' She smiled. 'Stand by, villagers. Ready, Ceridwen? Worm?'

There was a crackling noise through the public-address system as the vicar stepped up to the mike. 'Mr Mayor. Lady Mayoress. Ladies, gentlemen and children.' His voice echoed tinnily over the field. 'What a perfect day we are having to round off a truly memorable week.' He paused, smiling as a rumble of assent came from the crowd. 'We've been blessed with fine weather, not only today but all week. Each of our various events has gone splendidly and here we are, bathed in glorious sunshine and having the time of our lives.' More assent from his listeners.

'Let us not forget though, the reason for all this festivity. Let us remember whose heroism, whose martyrdom we celebrate here today.' A respectful silence settled over the field as Toby East spoke of how, exactly one thousand years ago, the village of Elsworth had been delivered from evil by the valour of its own dear saint, the maid Ceridwen, and of how this brave lass had later died a cruel death rather than renounce her faith. 'To remind us of these events,' he cried, 'and to bring to a climax this week of celebration, the children of Year Eight at Bottomtop Middle School now present their own production, entitled *Ceridwen – Heroine-Saint of Elsworth*.'

The vicar, with a sweeping gesture, indicated the marquee. There was a ripple of applause as Andrew Roberts emerged, followed by the villagers with Ceridwen in their midst. Andrew mounted the podium as the vicar vacated it. The villagers continued to the far end of the arena where a cluster of stalls and booths became the ancient village.

The narrator approached the mike. He was carrying his script, but in fact he was practically word-perfect without it. Vikings peeped from within the marquee as Andrew's voice rang out.

CHAPTER FORTY-NINE

' "The time – a little over one thousand years ago. The place – Elsworth, then a mere village, set in the midst of misty fenland. Elsworth, a once quiet village where terror now reigns, for the nearby fen has become the dwelling-place of a monster – a monster known to every terrified inhabitant as the Worm." '

An area behind the marquee was the fen. As Andrew paused in his narrative, the worm came capering round the side of the marquee and entered the arena. Gasps of admiration and surprise came from the crowd, but these became boos and hisses as the spectators entered into the spirit of the event. At the far end of the arena villagers cried out, pointing and scrambling to hide behind stalls and one another as the monster advanced. Fliss, who was not to appear till the worm had taken four victims, watched anxiously from behind a booth.

Everything went according to script. Gary's arms shot out and seized Tara Matejak, who screamed and writhed lustily as she was half carried, half dragged across the arena and away behind the marquee to the boos, whistles and catcalls of the crowd.

Michael Tostevin was the second victim. He threw away his mattock and tried to run, but the worm easily overcame him and he was borne away, howling, to join Tara.

When Haley Denton was seized, she managed to squeeze the contents of a sachet of ketchup all over her throat and chest. Cries of disgust and revulsion rose above the booing as she was dragged off, gurgling realistically and oozing gore.

Joanne O'Connor was to be victim number four. Fliss watched tensely as the girl moved on to the arena wielding a hoe, pretending to till the soil. Up to now everything was normal and Fliss felt a flicker of hope. She recalled what she'd said to Lisa on Thursday. '—that's where your game will have to stop because there's nothing special about the four of you—' At the time she hadn't been nearly as sure as she'd sounded, but now she dared to hope that she'd been right.

Joanne was working her way along a row of

imaginary carrots with her hoe. The worm was taking an unusually long time to appear and Fliss could see that Joanne was nervous. The poor girl didn't dare look towards the marquee because she wasn't supposed to see the worm approaching, but she was biting her lower lip and Fliss knew she just wanted her part to be over.

Fliss slitted her eyes against the glare of the sun and peered towards the marquee. As she did so, she heard a shrill scream and a figure appeared, running. It was Haley Denton, and she was followed by Tara Matejak and Michael Tostevin. Michael was trailing smoke and, as he pelted into the open, Fliss saw a flicker of flame and realized his tunic was on fire.

There were cries from people in the crowd. A woman ducked under the rope and sprinted towards the boy. She was carrying a car blanket. As Fliss watched, paralysed with shock, the woman brought Michael down with a rugby tackle and rolled him in the blanket. A man ran out to help her, but he'd got less than halfway when the marquee whooshed into flame and the worm came out of the smoke with fire in its jaws. The man cried out, skidded to a halt and ran back, scattering hysterical Vikings. The shrieking crowd milled as the monster blasted to left and right with jets of searing flame.

Andrew Roberts flung away his script, dived off the podium and vanished into the crush. Joanne

O'Connor abandoned her hoe and ran screeching towards where she thought she'd last seen her mother. An Army sergeant from the local recruiting office shouted to his five men to get into the armoured personnel carrier they'd been demonstrating. There was no live ammunition, but he thought that if they could ram the creature they might maim it. It was a bit of good thinking – one of the few bits to emerge in what was otherwise a shocked and panicky rabble – but it was to no avail. The first soldier was still some metres short of the A.P.C. when it was seized by two terrified civilians, who drove it off at speed.

The worm had advanced and stood now in the centre of the arena, snorting and clawing the turf with the talons of a gigantic bird. Its mad red eyes rolled this way and that and came to rest on the woman whose prompt action had saved the burning boy. She had flung her body across Michael's to prevent him rising. Now, as the monster swung its gaping maw her way, she cringed beside the heaving mound of blanket, helpless to save herself.

CHAPTER FIFTY

All of this had taken place in the space of a few seconds, during which Fliss, transfixed with horror, had seen all of her worst fears realized. Even as her brain was telling her such things were impossible, she knew that the Elsworth Worm had returned. Her four classmates, together with the contraption they had made, had undergone an incredible change to become the nightmare beast which now possessed the field. The beast which had burned the pigeon squabs and trampled the tulips. The beast which had rampaged through the supermarket a week ago, creating panic. The beast which was about to annihilate the woman who now crouched helpless in its path.

She didn't think. She was incapable of thought.

But as the worm prepared to blast its victim, she ran on to the field, waving her pathetic sword and shouting to attract the beast's attention.

The worm swung its great head, watching Fliss through hate-filled eyes. A long, low growl came out of its throat. Through the corner of her eye, Fliss saw her father leap the rope, heard him scream her name. She dashed on. It was as though some force had assumed control of her mind, of her actions. She felt no fear now.

With a shattering roar, the worm launched itself to meet her, discharging a shaft of flame which passed so close she felt its searing heat. As they met in midfield she swung the sword at the creature's scaly neck, but it glanced off as though the beast were clad in steel. The monster dwarfed her as it reared to rip with its claws. She ducked and dodged as razor talons flailed the air. The flimsy sword windmilled around her head till, inevitably, it struck a horny claw and was torn from her grip.

It's over now, she thought. It must be. One slash of those talons and I'll fall in shreds. And even as she thought this, a voice in her head was crying, 'Forward. Only forward.' She pressed on without knowing why, ducking and weaving. So close was she now to her adversary that the worm could neither see her nor bring its fire to bear, and every time it backed up to get her range, Fliss moved with it.

It couldn't continue, and Fliss knew it. She felt herself tiring. Dimly, she was aware that the Festival Field was emptying as townspeople scrambled over walls and fences or fought their way through gateways. Perhaps, she thought, some will escape if they flee, but the worm will have its revenge on Elsworth, and slake its thousand-year hunger with Elsworth's dead. She wished her parents would save themselves, but knew they were near for her sake. Her limbs felt leaden and she couldn't get her breath. She knew that soon she must fall.

It happened almost at once. The worm backed up and, as Fliss followed, her sandal came down on a stone. She stumbled and fell, and before she could roll or rise, she was pinned to the turf by a great taloned foot. She gritted her teeth and screwed up her eyes, awaiting the blast which would finish her.

It didn't come. Instead, the worm emitted a chilling screech and the foot was snatched back. Fliss rolled and looked up. There stood the beast, but as she watched, its image began to shimmer and warp like an object underwater. She screwed up her eyes and shook her head. It was changing, shrinking. The coils of smoke, the jets of flame, became wisps and tongues which flickered out and dispersed before her eyes. The scaly armour seemed to soften and hang in folds and wrinkles, and the creature's sinewy limbs disintegrated, becoming thin and pale as the talons in which they ended curled and shrivelled like feathers in a flame. The screeching roar dwindled through cough, bark and groan till it resolved in the anguished cries of children.

A wave of nausea swept through Fliss and she closed her eyes. When she looked again the worm had gone. On the scorched and trampled grass lay a smashed thing – a contraption of wood and cloth and wire in the midst of which sprawled four ashen-faced children. A hand plucked at the sleeve of her dress. Fliss turned. The woman she'd saved gazed into her eyes. 'What – what was it?' she croaked. 'What happened?'

Fliss shook her head. She felt unutterably tired. 'I don't know,' she murmured. 'But whatever it was, I think it's over now.'

CHAPTER FIFTY-ONE

A week went by before life in Elsworth returned to something like normal. During that time, two explanations emerged for what had taken place on the Festival Field.

The vicar said that Elsworth had once more been threatened, and once more delivered.

The *Star* abandoned its earlier sensationalism and said that the townspeople had been the victims of a collective hallucination, and of mass hysteria.

The town's churchgoers tended to favour the vicar's version, while the police and most other people went along with the *Star*. No prosecutions followed the recent spate of vandalism. Nobody felt like delving any deeper into the matter for fear of uncovering fresh mysteries. No. It was over and done with, whatever it was. Forget it. Life goes on.

Fliss could not forget it, and neither could Lisa, Ellie-May, Gary or Trot. They'd survived, but their horrific experience had left them feeling isolated – set apart somehow from the world of friends, family and everyday life. Saturday found them huddled in the greenhouse on the abandoned allotment. The spell of fine weather had broken down. Rain hissed and rattled on the grimy panes, there was no sign of Hughie Ackroyd, and they were glad of the warmth which came from the rusty stove.

They'd sat for some time in silence, letting a chill which had little to do with the weather thaw from their bones, when Lisa said, 'I don't know how you can stand to be with us, Fliss, after what we did to you.'

Fliss shook her head 'It wasn't you, Lisa. It wasn't any of you. You were possessed – taken over by something. It started as soon as you were chosen to play the worm. It had waited a thousand years and it didn't rush. It took over your minds, little by little. Then it started changing your bodies, though you didn't know it. On Festival Day, behind that marquee, it extinguished you altogether and became itself once more –

the Elsworth Worm, bent on revenge. If others had been chosen, the same would have happened to them.'

'I know.' Ellie-May shivered. 'I could feel it. It was like – you know – you get an urge to do something you know's wrong, but the thought of it's so exciting you can't stop yourself. It was terrific and horrible, both at the same time.'

'I felt like that,' nodded Trot. 'I wanted to do the worst things I could think of, even though they were stupid and cruel. I just couldn't help it.'

'I knew something was happening,' murmured Gary. 'Deep down I knew, but I didn't want to admit it to myself. I was enjoying it all, you see. The power. People's fear. The destruction.'

'What I want to know,' said Fliss, 'is what it felt like when you actually became the worm. I mean, did you know you'd changed?'

Lisa shook her head. 'There was this terrific excitement, that's all. You felt like you could do anything. Anything at all. Everybody was scared of you, see? It gave you power – a feeling of power.'

'And hate,' put in Gary. 'You hated every-body and everything. You just wanted to smash everything in sight.' He grinned ruefully, shaking his head. 'You should've felt the hate we felt for

189

you, Fliss. You and your plastic sword. It was awesome.'

'I felt it,' said Fliss.

'But you came on,' said Lisa. 'I wonder what would have happened if you hadn't?'

'We'd have become murderers,' said Trot. 'The four of us. We just wanted to destroy everyone and everything in Elsworth. We must've been totally crazy.'

Fliss shook her head. 'I told you, Trot – it wasn't you.'

'But if it hadn't been for you, Fliss, the worm would've won and we wouldn't exist as separate people – or as people at all, come to that.'

Fliss shook her head again. She smiled, her first smile in a long time.

'It wasn't me either,' she said.

ABOUT THE AUTHOR

Robert Swindells left school at fifteen and worked as a copyholder on a local newspaper. At seventeen he joined the RAF for three years, two of which he served in Germany. He then worked as a clerk, an engineer and a printer before training and working as a teacher. He is now a full-time writer and lives on the Yorkshire moors.

He has written many books for young readers, including *Room 13*, the winner of the 1990 Children's Book Award, *Dracula's Castle*, *The Postbox Mystery*, *Hydra* and *The Thousand Eyes of Night*, all of which are available in Yearling paperback. His books for older readers include the award-winning *Brother in the Land*. As well as writing, Robert Swindells enjoys keeping fit, travelling and reading.

A SELECTED LIST OF TITLES
AVAILABLE FROM YEARLING BOOKS

THE PRICES SHOWN BELOW WERE CORRECT AT THE TIME OF GOING TO PRESS. HOWEVER TRANSWORLD PUBLISHERS RESERVE THE RIGHT TO SHOW NEW RETAIL PRICES ON COVERS WHICH MAY DIFFER FROM THOSE PREVIOUSLY ADVERTISED IN THE TEXT OR ELSEWHERE.

☐	86297 3	**THE DEMON PIANO**	*Rachel Dixon*	£2.99
☐	86299 X	**THE WITCH'S RING**	*Rachel Dixon*	£2.99
☐	86277 9	**SHRUBBERY SKULDUGGERY**	*Rebecca Lisle*	£2.99
☐	86325 2	**THE WEATHERSTONE ELEVEN**	*Rebecca Lisle*	£2.99
☐	86227 2	**ROOM 13**	*Robert Swindells*	£2.99
☐	86275 2	**THE POSTBOX MYSTERY**	*Robert Swindells*	£2.50
☐	86278 7	**DRACULA'S CASTLE**	*Robert Swindells*	£2.50
☐	86313 9	**HYDRA**	*Robert Swindells*	£2.99
☐	86316 3	**THE THOUSAND EYES OF NIGHT**	*Robert Swindells*	£2.99
☐	86201 9	**THE CREATURE IN THE DARK**	*Robert Westall*	£2.50

All Yearling Books are available at your bookshop or newsagent, or can be ordered from the following address:
Transworld Publishers Ltd,
Cash Sales Department,
P.O. Box 11, Falmouth, Cornwall TR10 9EN

Please send a cheque or postal order (no currency) and allow £1.00 for postage and packing for one book, an additional 50p for a second book, and an additional 30p for each subsequent book ordered to a maximum charge of £3.00 if ordering seven or more books.

Overseas customers, including Eire, please allow £2.00 for postage and packing for the first book, an additional £1.00 for a second book, and an additional 50p for each subsequent title ordered.

NAME (Block Letters) ....

ADDRESS ....

....